T0291168

INTEGRATION OF ENERGY, INFORMATION, TRANSPORTATION AND HUMANITY

INTEGRATION OF ENERGY, INFORMATION, TRANSPORTATION AND HUMANITY
Renaissance from Digitization

C.C. CHAN
The University of Hong Kong, Hong Kong SAR, P.R. China

GEORGE YOU ZHOU
National Institute of Clean and Low Carbon Energy, Beijing, P.R. China

WEI HAN
Hong Kong University of Science & Technology, Guangzhou Campus (HKUST Guangzhou) Guangzhou, Guangdong Province, P.R. China

ELSEVIER

Elsevier
Radarweg 29, PO Box 211, 1000 AE Amsterdam, Netherlands
The Boulevard, Langford Lane, Kidlington, Oxford OX5 1GB, United Kingdom
50 Hampshire Street, 5th Floor, Cambridge, MA 02139, United States

Notices

Knowledge and best practice in this field are constantly changing. As new research and experience broaden our understanding, changes in research methods, professional practices, or medical treatment may become necessary.

Practitioners and researchers must always rely on their own experience and knowledge in evaluating and using any information, methods, compounds, or experiments described herein. In using such information or methods they should be mindful of their own safety and the safety of others, including parties for whom they have a professional responsibility.

To the fullest extent of the law, neither the Publisher nor the authors, contributors, or editors, assume any liability for any injury and/or damage to persons or property as a matter of products liability, negligence or otherwise, or from any use or operation of any methods, products, instructions, or ideas contained in the material herein.

ISBN: 978-0-323-95521-8

For Information on all Elsevier publications
visit our website at https://www.elsevier.com/books-and-journals

Publisher: Joseph P. Hayton
Acquisitions Editor: Kathryn Eryilmaz
Editorial Project Manager: Aleksandra Packowska
Production Project Manager: Jayadivya Saiprasad
Cover Designer: Victoria Pearson Esser

Typeset by MPS Limited, Chennai, India

Working together
to grow libraries in
developing countries

www.elsevier.com • www.bookaid.org

Contents

List of contributors   *xi*

Preface   *xv*

Acknowledgments   *xxi*

Introduction   *xiii*

1. Challenge and trend on energy digitalization   1

Wei Han, C.C. Chan, Youhao Hu, Chang Liu and George You Zhou

1.1 Energy security from artificial intelligence—driven energy   1
    1.1.1 The brief review of artificial intelligence   1
    1.1.2 The artificial intelligence in energy systems   3
    1.1.3 The security of artificial intelligence—driven energy   7
1.2 Carbon neutrality from energy ecosystem   11
    1.2.1 Natural and energy ecosystem   11
    1.2.2 Energy ecosystem roadmap   15
    1.2.3 Carbon neutrality in energy ecosystem   16
1.3 From industrial revolution 4.0 to 5.0   18
    1.3.1 From Industry 1.0 to Industry 4.0   18
    1.3.2 The Industry 5.0   21
    1.3.3 Technology enablers of Industry 5.0   23
1.4 Emerging sign of supersmart Society 5.0   26
    1.4.1 What is Society 5.0?   26
    1.4.2 The evolutionary direction of society   29
    1.4.3 Realizing Society 5.0   33
References   36

2. Human-centered Renaissance from energy digitization   39

Chunhua Liu, C.C. Chan, Kuo Feng and George You Zhou

2.1 Renaissance phenomena: cyber-physical-social   39
    2.1.1 Concept of industrial humanity   39
    2.1.2 Development of humanities network   41
    2.1.3 Integration key of cyber-physical-social   43
    2.1.4 Integration applications of humanistic network   45
2.2 Renaissance challenge: human-artificial intelligence-environment   47
    2.2.1 Background   47
    2.2.2 Challenges of artificial intelligence in the Renaissance   52

	2.2.3	Industrial humanity in the Renaissance	56
2.3		Renaissance elements: human-machine-things	57
	2.3.1	The definition of human-machine-thing	58
	2.3.2	The importance of the human network in human-machine-thing	59
	2.3.3	The network integration	60
	2.3.4	Conclusion	64
2.4		Renaissance foundation: philosophy-science-engineering	64
	2.4.1	Interactively integrating philosophical	65
	2.4.2	Scientific theory	67
	2.4.3	Engineering practice	69
References			71

3. Integration of energy, transportation, and information with humanity   **73**

Xiaohua Li, C.C. Chan, Hang Zhao, Jin Li, Yin Yao and George You Zhou

3.1		Social behavior patterns under the information revolution	73
	3.1.1	Development history and process of the information revolution	73
	3.1.2	Main technologies of the contemporary information revolution	76
	3.1.3	Integration and application of information network technology and the future Metaverse	81
3.2		The energy revolution solves climate problems: carbon peaking and carbon neutrality	83
	3.2.1	The energy revolution triggered by the climate problem	83
	3.2.2	Opportunities and challenges under the energy revolution	88
	3.2.3	Development of energy revolution under the 4N4F	93
	3.2.4	Typical cases of 4N4F in the energy revolution	98
3.3		Changes in mobility patterns brought about by the transportation revolution	99
	3.3.1	Transportation reform in the contemporary era	99
	3.3.2	Changes in the transportation industry brought about by the 4N4F	101
	3.3.3	Specific changes brought about by the change in transport pattern	107
3.4		People oriented for four networks for fusion	111
	3.4.1	People-oriented Industry 5.0	111
	3.4.2	Digital humanity network and its application in 4N4F	115
	3.4.3	The 4N4F drives the transformation from Industry 4.0 to Industry 5.0	120
References			122

4. Strategy of integrated humanity for specific applications   125

Chaoqiang Jiang, C.C. Chan, Tianlu Ma, Jingchun Xiang, Xiaosheng Wang,
Chen Chen, Hongbin Sun and George You Zhou

4.1   Energy Internet initiative   125
   4.1.1   Energy Internet summary   125
   4.1.2   Energy flow and information flow for smart automobile charging   128
   4.1.3   Smart energy system   130
   4.1.4   Integrated energy system   133
4.2   Intelligence connected transportation   137
   4.2.1   Electrification, enabling green travel   137
   4.2.2   Intelligence connected transportation   141
   4.2.3   Smart transportation system   144
4.3   Collaborative robots for industrial Internet-of-Things   147
   4.3.1   Industrial Internet-of-Things   147
   4.3.2   Collaborative robots in Internet-of-Things   149
   4.3.3   Smart industry system   152
4.4   Ecosystem operating system for smart city   156
   4.4.1   Ecosystem in smart city   156
   4.4.2   Low carbon smart city   159
   4.4.3   Smart city with intelligent computing   161
References   165

5. Framework of Digital Renaissance with human-in-the-loop   169

Christopher H.T. Lee, C.C. Chan, Yaojie He, Chenhao Zhao,
Huanzhi Wang and George You Zhou

5.1   Design pattern from first principle   169
5.2   Data-driven cyber-physical Digital Twin   171
   5.2.1   Background   171
   5.2.2   Concept of Digital Twin   173
   5.2.3   Model of Digital Twin   175
   5.2.4   Application case study and challenge of Digital Twin   177
5.3   Knowledge-driven human-cyber Hybrid Twin   180
   5.3.1   Background   180
   5.3.2   Concept and challenges of Virtual Twin   181
   5.3.3   Concept and model of Hybrid Twin   183
   5.3.4   Application of Hybrid Twin   186
5.4   Intelligence-driven human-cyber-physical Cognitive Twin   188
   5.4.1   Background   188
   5.4.2   Characteristics and structure of Cognitive Twin   189

	5.4.3	Enabling technology for model management	191
	5.4.4	Function layers of Cognitive Twin	193
	5.4.5	Applications and challenges of Cognitive Twin	195
	5.4.6	Future milestones	196
References			200

6. Examples of energy, transportation, and information with humanity   203

Wei Han, C.C. Chan, George You Zhou, Zhiyong Yuan, Yingjie Tan,
Hong Rao, Tik Lou, Jiawei Wu, Haohong Shi, Anjian Zhou, Changhong Du,
Guocheng Lu, Yue Qiu, Suyang Zhou, Wei Zhang, Ying Li,
Chunying Huang, Hailong Cheng, Mingxu Lei, Dan Tong and Chi Li

6.1	Build the new type power system with new energy as the main body		204
	6.1.1	Main features of the new type power system	205
	6.1.2	Empowering the new type power system with digital power grid	206
	6.1.3	Intelligent integrated energy services	208
	6.1.4	The basic theory and methods for intelligent integrated energy services	210
	6.1.5	Typical intelligent integrated energy services platforms	211
6.2	Behavior-driven smart vehicle and green transportation		212
	6.2.1	Introduction	213
	6.2.2	Future trends of electric vehicles	215
	6.2.3	The development of smart vehicles	220
	6.2.4	Conclusion	228
6.3	Co-robot-driven smart manufacturing plant and process		229
	6.3.1	Human—robot collaboration	229
	6.3.2	Discussion	235
	6.3.3	The future of cobot	235
	6.3.4	Cobots in the age of Industry 5.0	237
	6.3.5	Conclusion	238
6.4	Intelligent integrated energy services and society ecosystem		238
	6.4.1	Integrated energy system and intelligent integrated energy services	238
	6.4.2	Basic characteristics of intelligent integrated energy services	240
	6.4.3	The basic theory and methods for intelligent integrated energy services	242
	6.4.4	Typical intelligent integrated energy services platforms	243
6.5	Digital Renaissance-driven smart city		245
	6.5.1	The emergence of smart city	247
	6.5.2	Smart city: the integration and innovation of four network four flows	254

    6.5.3 Cases and practices   257

    6.5.4 Conclusion   262

6.6 Practice of four network four flows integration: an approach to
digital Renaissance   262

    6.6.1 Four network four flows visualization: promotion to cyber and
physical worlds integration   263

    6.6.2 Four network four flows value system: promotion to cyber and
humanistic worlds integration   271

    6.6.3 Four network four flows—based smart city: a comprehensive
integration of physical, humanistic, and cyber worlds   276

    6.6.4 Four network four flows practice: an approach to great digital
Renaissance   283

    6.6.5 Conclusion   286

References   287

index   *291*

List of contributors

C.C. Chan
The University of Hong Kong, Hong Kong SAR, P.R. China

Chen Chen
Department of Electrical Engineering, City University of Hong Kong, Hong Kong

Hailong Cheng
Institute of Digital Guangdong, Guangzhou, P.R. China

Changhong Du
Deepal Automobile Technology Co., Ltd., P.R. China

Kuo Feng
School of Energy and Environment, City University of Hong Kong, Hong Kong

Wei Han
Sustainable Energy and Environment Thrust, The Hong Kong University of Science and Technology (Guangzhou), Guangzhou, P.R. China

Yaojie He
School of Electrical and Electronic Engineering, Nanyang Technological University, Singapore

Youhao Hu
Sustainable Energy and Environment Thrust, The Hong Kong University of Science and Technology (Guangzhou), Guangzhou, P.R. China

Chunying Huang
Institute of Digital Guangdong, Guangzhou, P.R. China

Chaoqiang Jiang
Department of Electrical Engineering, City University of Hong Kong, Hong Kong

Christopher H.T. Lee
School of Electrical and Electronic Engineering, Nanyang Technological University, Singapore

Mingxu Lei
Institute of Digital Guangdong, Guangzhou, P.R. China

Chi Li
International Academicians Science & Technology Innovation Center, P.R. China

Jin Li
Electric Power Engineering, Shanghai University of Electric Power, Shanghai, P.R. China

Xiaohua Li
Electric Power Engineering, Shanghai University of Electric Power, Shanghai, P.R. China

Ying Li
Institute of Digital Guangdong, Guangzhou, P.R. China

Chang Liu
Sustainable Energy and Environment Thrust, The Hong Kong University of Science and Technology (Guangzhou), Guangzhou, P.R. China

Chunhua Liu
School of Energy and Environment, City University of Hong Kong, Hong Kong

Tik Lou
XEV Company, Italy

Guocheng Lu
Deepal Automobile Technology Co., Ltd., P.R. China

Tianlu Ma
Department of Electrical Engineering, City University of Hong Kong, Hong Kong

Yue Qiu
Southeastern University, Nanjing, P.R. China

Hong Rao
China Southern Power Grid Co., Ltd. (CSG), P.R. China

Haohong Shi
XEV Company, Italy

Hongbin Sun
Department of Electrical Engineering, Tsinghua University, Beijing, P.R. China; College of Electrical and Power Engineering, Taiyuan University of Technology, Taiyuan, P.R. China

Yingjie Tan
China Southern Power Grid Co., Ltd. (CSG), P.R. China

Dan Tong
Institute of Digital Guangdong, Guangzhou, P.R. China

Huanzhi Wang
School of Electrical and Electronic Engineering, Nanyang Technological University, Singapore

Xiaosheng Wang
Department of Electrical Engineering, City University of Hong Kong, Hong Kong

Jiawei Wu
XEV Company, Italy

Jingchun Xiang
Department of Electrical Engineering, City University of Hong Kong, Hong Kong

Yin Yao
Electric Power Engineering, Shanghai University of Electric Power, Shanghai, P.R. China

Zhiyong Yuan
China Southern Power Grid Co., Ltd. (CSG), P.R. China

Wei Zhang
State Grid Integrated Energy Service Group Co., Ltd., P.R. China

Chenhao Zhao
School of Electrical and Electronic Engineering, Nanyang Technological University, Singapore

Hang Zhao
Robotics and Autonomous Systems Thrust, The Hong Kong University of Science and Technology (Guangzhou), Guangzhou, P.R. China

Anjian Zhou
Deepal Automobile Technology Co., Ltd., P.R. China

George You Zhou
National Institute of Clean and Low-Carbon Energy, Beijing, P.R. China

Suyang Zhou
Southeastern University, Nanjing, P.R. China

Preface

The mission of scientists is to seek the truth, reveal the laws of nature, and benefit all mankind. The spirit of science is freedom of inquiry and independence. The scientific methodology is logical reasoning, deductive calculation, and experimental verification. The question to be answered by science is "why?". Engineering is the integration of science, technology, and management to solve real-world problems and make the world a better place. The question to be answered by engineering is "how?". Technology is skill, craftsmanship, to pursue excellence. This book upholds these principles to address the sustainable development goals, carbon neutrality, and harmonious coexistence between human kind and nature by the integration of energy, transportation, and information with humanity, in the context of the renaissance of digitation and the deep integration of human, cyber, and physical worlds.

Digital productivity, a typical feature of the Fourth Industrial Revolution that distinguishes it from the previous three industrial revolutions, needs to be supported by the economic foundation and superstructure. The energy, information, and transportation networks are the three pillars of the economic foundation, while the humanistic network is an important part of the superstructure. Through the integration of **Four Networks** (energy, information, transportation, and humanity) and **Four Flows** (energy, information, material, and value), the proactive initiative of humanity can combine the energy, information, and transportation (mobility) revolutions, to establish a new production relationship formed within the "human−cyber−physical" system. This integration can further extract a huge productivity increase when it has benefit from the big data approach utilized in the Fourth Industrial Revolution. Such exponentially growing productivity is basically derived from the sum of the previous productivities of the past three industrial revolutions.

The "Four Networks and Four Flows (**4N4F**)" represents the fundamental changes and impacts of human thinking in contribution to the evolution from the Fourth to the Fifth Industrial Revolution. The core of the Fourth Industrial Revolution is artificial intelligence (AI), whereas the cores of Fifth Industrial Revolution are humanity and the environment in addition to AI. This puts forward a strategic fusion need for **4N4F** that integrates philosophical thinking, scientific theory, and engineering

practice, as well as mastering the three revolutions, namely, energy, information, and a transportation (mobility). In the philosophical perspective,
it is essential to establish holistic thinking, based on the interaction
between the economic foundation and superstructure, to create an advantage that the whole of the fusion is greater than the sum of the individual
parts. From scientific theory, it is essential to explore the laws of interaction between energy, information, and human behavior to turn waste
energy into useful energy and promote carbon neutrality. In terms of
engineering practice, it is essential to combine energy technology and
information technology, develop a smart energy operating system, adopt
an End-Edge-Cloud structure, and utilize energy systems with AI/big
data systems to achieve value-added benefits.

At present, the "new infrastructure" has received widespread attention,
focusing on the amalgamation of "smart energy" and "smart transportation." Specifically, it involves the technological progress and integration
of multiple innovative fields such as the mass application of electric vehicles and their charging stations, photovoltaic buildings, distributed energy
storage, hydrogen energy, edge computing, the Internet-of-Things,
blockchain, metaverse, and generative AI. Basically, its essence can be
defined and planned uniformly through the technical concept of **4N4F**.

The development of the strategic **4N4F** framework will accelerate the
application of the Internet-of-Things and AI technologies in the energy,
information, and transportation industries. It will further promote deep
cross-border integration in the process of technological innovation and
facilitate the realization of open sharing platforms of intelligent industry
collaboration. The integration of energy and information networks,
together with the integration of distributed energy and distributed supercomputing, can form many new growth areas, including green industry
and ecological chain development, and incubate new technologies, industries, and business models. Fusion is a deep amalgamation that produces
long-term benefits as the value of the whole is greater than the sum of
the parts.

The far-reaching significance of intelligent connected electric vehicles
is from smart vehicles to Internet of Vehicles, to smart transportation, to
smart city, and to smart society. Our overall goal is people oriented, harmonious coexistence between man and nature, and sustainable development. The integration of 4N4F is the foundation. Its fundamental theory
is deep integration of the human world, the information world or cyber
world, and the physical world.

In short, the objectives of the integration of 4N4F; the philosophy, fundamental theory, science, and engineering; the mechanism, keys, and benefits; the carriers; and the framework are summarized in the following paragraphs.

The objective of the integration of 4N4F is to give full play to the advantages of digital technology and digital economy in an all-round way.

The philosophy—whole is greater than the sum of each part; the economic foundation interacts with the superstructure. **The fundamental theory**—integration of the human world, the information world or cyber world, and the physical world. **The science theory**—exploring the fundamental relationships amongst energy, information, and human behavior. **The engineering**—combining energy and information technology through a smart energy operating system, with energy technology combined with cloud technology, edge computation, AI, and big data technology, to achieve value-added results.

The mechanism of the integration of 4N4F is essentially via coupling and sharing, increase utilization of energy and assets. **The keys** of the integration of 4N4F lies in and is via the humanity network, revealing dynamic causality and complex logic. **The core benefit** of the integration of 4N4F is to gain greater economic and environmental benefits.

The carriers of the integration of 4N4F are mainly (1) integrated energy intelligent management systems, (2) intelligent connected electric vehicles which are mobile carriers, and (3) intelligent connected building-integrated photovoltaics buildings which are standstill carriers. This also includes smart renewable energy, distributed green energy, smart transportation, smart city, and smart society.

The framework of the integration of 4N4F is fundamentally a 5G-based fusion network, an energy-based industrial fusion, and a digital economy-based ecological fusion. The main integration elements are data, algorithms, computing power, and platforms.

We must deeply integrate the human world, the information world, and the physical world. A digital twin is a digital representation of a physical world, that is, the integration of the physical and information worlds, to convert data to information. But the digital twin is not enough by itself, we should further transfer information into knowledge by using a so-called hybrid twin with the help of the humanity world. Furthermore, with the help of the humanity world, we should further transfer knowledge into intelligence by so-called cognitive twin, which constantly updates knowledge and is autonomous. In this way, through the deep integration of the human world, information world, and physical world, we are able to turn data into information, information into knowledge, and knowledge into intelligence Thus, to integrate Einstein's energy law and Shannon's information law/entropy law, to discover the correlation among energy, information, and human behavior. Such as to convert waste energy which is disorder to useful energy which is in order, to achieve greater economic and environmental benefits.

The Fourth and Fifth Industrial Revolutions will fundamentally change our lives as well as the lives of future generations and recreate the economy, society, culture, and environment on which mankind depends for survival. We need to think disruptively and strategically to deal with the Fourth and Fifth Industrial Revolutions. The integration of 4N4F is to replace the traditional linear thinking with a breakthrough circular comprehensive thinking. It links the energy, information, and transportation (mobility) revolutions and formulates the humanity network of the superstructure with the economic integration of the energy, information, and transportation networks, resulting in disruptive economic and ecological benefits. As the inventor of the theory and practice of "Integration of Four Networks and Four Flows," I hope the publication of this book can further advocate the formation and development of the new industrial ecology of the "Integration of Four Networks and Four Flows" and promote the new infrastructure strategy. We must seize the opportunity to be at the forefront of the Fourth and Fifth Industrial Revolutions and the development of the digital economy.

In the last century, our focus has been on the advancement of the physical world. At the beginning of this century, we focused on the development of the cyber world and the revival of digitalization. However, we should further develop our disruptive thinking, deeply integrating the physical, cyber, and humanity worlds, effectively transforming data into information, information into knowledge, and knowledge into intelligence to solve complex problems. In the real world, technology alone cannot solve complex issues and meet new challenges. The human world, the cyber world, and the physical world must also be deeply integrated.

The history of human development marches from agriculture economy to industrial economy to currently digital economy. The energy transitions from wood to coal, from coal to oil and gas, and from fossil energy to currently renewable energy. We are now at the milestone of transition. At this junction, who can identify the key issues and problems, and give the total solution, will be the leader in the forefront.

It is really exciting that we are engaging in a program that would have major impacts on the future of our society and the welfare of our future generations.

C.C. Chan

Acknowledgments

Thoughts and materials presented in this book are the collection of the research, teaching, and industrial practice of Prof. C.C. Chan and his colleagues over half century. It is our great pleasure to present this book in celebration of the foundation of the International Academicians Science & Technology Innovation Centre and the 86th anniversary of Prof. C.C. Chan.

We are deeply grateful to Professor J.F. Eastham who wisely reviewed and polished the book. We are deeply indebted to our friends and colleagues globally for their encouragement and support. We highly appreciate the editors of Elsevier Publishing Company for their effective support and who made this book possible. And of course, we thank our families for their unconditional support and absolute understanding during the writing of this book.

Introduction

Focusing on energy, transportation, information, and economic networks and flows, *Integration of Energy, Information, Transportation and Humanity* uniquely examines the interconnection, interaction, and integration across these multiple sectors. It helps readers understand the correlation of energy, transportation, and information via the integration of humanistic, cyber, and physical worlds. It clearly explains the objectives of the integration of energy, transportation, information, and humanity networks, as well as the integration of energy, information, material, and value flows (4N4F); the philosophy, science, and engineering of the integration of 4N4F; the mechanism, keys, and benefits of the integration of 4N4F; the carriers of the integration of 4N4F; and the framework of the integration of 4N4F.

The key features of the book include (1) the syntheses of the newest developments in digital technologies and the digital economy, (2) case studies and examples that illustrate the application of the methodologies and technologies employed, and (3) it is useful for both theoretically and technically oriented researchers.

Chapter 1 lays down the background of digital technologies permeate modern life, affecting everything from the way we work and travel to the way we live and play. Energy digitalization promises to help improve the safety, efficiency, sustainability, and productivity of the global energy system. The digital energy system of the future may be able to identify who needs energy and deliver it to the right place, at the right time, and at the lowest cost. Digitalization is not only improving the flexibility, robustness, and accessibility of energy systems but also raising attention to the security from artificial intelligence (AI)-driven energy, carbon neutrality from energy eco-systems, the transition from the industrial revolutions 4.0 to 5.0, the emerging signs of super smart society 5.0, and so on. This chapter seeks to provide readers with a clearer understanding of what digitalization means for energy—shining a light on both its enormous potential and its most pressing challenges.

Chapter 2 reviews that before the third industrial revolution, humans mainly controlled the operation of energy systems. While currently with the development of energy digitization, big data and AI algorithms are widely applied to control the systems. People are liberated from labor, but the purpose of people orientation should not be ignored. In the fourth industrial revolution in the future, the network of humanities will be revived as the superstructure, forming a new "four networks and four flows." This chapter mainly introduces the phenomena, challenge, elements, and foundations of human-centered renaissance from energy digitization.

Chapter 3 discusses the new stage of the sixth information revolution, the concept of "four networks and four flows" is applied based on the online human−computer interaction networks. This chapter presents an overview of the progression of the information revolution and its amalgamation with the forthcoming Metaverse. Initially, the potential opportunities and challenges associated with the energy revolution are proposed. Subsequently, the transportation transformation during the era of humanistic digital transformation is deliberated upon. Finally, the pivotal role of the human network in intelligent applications is underscored.

Chapter 4 describes that the core of the Four Networks Integration is the integration of the humanistic network under the "human−cyber−physical" ecosystem. This embodies the top-level design thinking for the development of the digital economy under the background of the new infrastructure, that is, the method of establishing the integrated development of human−cyber−physical systems. This is reflected in the applications of smart energy, smart industry, smart transportation, and smart cities. Specifically, in this chapter, the smart energy system including energy flow and information is discussed; the smart transportation system relying on communications and the exchange of data between different platforms will be investigated; the collaborative robots for industrial IoT in the smart industry system will be explored; and the eco-system operating system for smart city with intelligent computing will be investigated. It is shown that the integration of the humanistic network has become the key to accelerating different intelligent applications.

Chapter 5 shows that data communication between physical entities and virtual entities enables an innovative method for industry to organize the product lifecycle. In addition, the humanistic knowledge and information from the customer side should be brought in, since they play a crucial role in the intelligent industry and smart society. Hence, the data, information, and knowledge interactions among the social, physical, and cyber worlds should be modeled within a dynamic system. With the human−physical−cyber system, intelligence cognition is achieved with the ability to perform prediction, synchronization, and optimization in the digital ecology. This chapter provides the detailed elements and structure of the human—physical—cyber system, as well as the evolution, milestones, and challenges in building it.

Chapter 6 summarizes that the communication of data between physical and virtual entities has revolutionized the way the industry organizes the product lifecycle. By leveraging the power of data, companies can optimize their operations, reduce costs, and improve their products and services. However, data alone is not enough. To achieve true intelligence in the industry and society, it is essential to incorporate humanistic knowledge and customer information. After all, customers are the ultimate arbiters of success in any industry, and their needs and preferences must be considered to achieve sustainable growth. Overall, the human—physical—cyber system represents a powerful tool for achieving intelligence in industry and society. By leveraging the power of data and incorporating humanistic knowledge and customer information, we can build a smarter, more sustainable future for all. Case studies in the following fields are discussed: (1) building new power system with new energy as the main body, (2) smart vehicles and green transportation, (3) smart manufacturing, (4) intelligent integrated energy services and society ecosystem, (5) digital renaissance—driven smart city, and (6) practice of integration of four networks four flows with visualization.

The core of far-reaching integration of energy, information, transportation and humanity lies on people-oriented, harmonious coexistence between humans and nature, together with sustainable development, so that both economic development and environment protection are achieved. The basic foundation principle is the deep integration of the humanistic, cyber, and physical worlds. We are excited to be engaged in the endeavors that will have significant impact on the welfare of our future generations.

CHAPTER ONE

Challenge and trend on energy digitalization

Wei Han[1], C.C. Chan[2], Youhao Hu[1], Chang Liu[1] and George You Zhou[3]

[1]Sustainable Energy and Environment Thrust, The Hong Kong University of Science and Technology (Guangzhou), Guangzhou, P.R. China
[2]The University of Hong Kong, Hong Kong SAR, P.R. China
[3]National Institute of Clean and Low-Carbon Energy, Beijing, P.R. China

Abstract

Digital technologies permeate modern life, affecting everything from the way we work and travel, to the way we live and play. Energy digitalization promises to help improve the safety, efficiency, sustainability, and productivity of the global energy system. The digital energy system of the future may be able to identify who needs energy and deliver it to the right place, at the right time, and at the lowest cost. Digitalization is not only improving the flexibility, robustness, and accessibility of energy systems, but also raising attention to the security from artificial intelligence (AI)—driven energy, carbon neutrality from energy ecosystems, the transition from the industrial revolutions 4.0 to 5.0, the emerging signs of supersmart society 5.0, and so on. This chapter seeks to provide readers with a clearer understanding of what digitalization means for energy—shining a light on both its enormous potential and its most pressing challenges.

Keywords: Energy digitalization; energy security; carbon neutrality; industrial revolution

1.1 Energy security from artificial intelligence—driven energy

1.1.1 The brief review of artificial intelligence

As illustrated in Fig. 1—1, the research in artificial intelligence (AI) has experienced three main stages of development. From the mid-1950s to the 1960s, the first stage presented the AI concept, mainly focusing on machine translation of logical reasoning. The symbolism developed rapidly, and expert systems and knowledge engineering became the research

Integration of Energy, Information, Transportation and Humanity.
DOI: https://doi.org/10.1016/B978-0-323-95521-8.00006-3

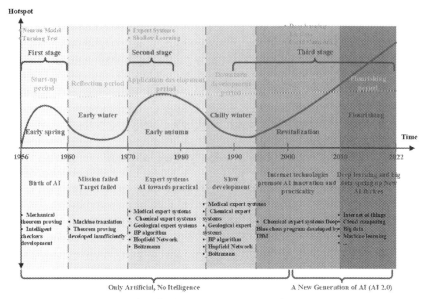

Figure 1−1 The process of artificial intelligence (AI) evolution.

mainstream. After 1960, AI research went through several ups and downs. At this stage, Herbert Simon and Allen Newel developed an automatic theorem-proving system called the Logic Theorist, which was subsequently utilized to prove all theorems in the book Principia Mathematica in less than 2 months. People gradually realized that logical reasoning alone was insufficient to reach the level of machine intelligence. Thus the AI research spontaneously entered the second stage.

From the 1970s to the 1980s, the second stage of AI focused on how to summarize knowledge and then hand over to a computer system for processing. These systems primarily seek to summarize the problem-solving knowledge of human experts, consequently, such knowledge is programmed into computer systems to produce "expert systems" that can be used to solve real-world problems. Nevertheless, such investigations have been slow to progress due to high development costs. As a result, AI development entered another trough. At this stage, researchers hoped to model and mine the tacit knowledge in the dataset through machine learning (ML) techniques. Hence, the mainstream research of AI stepped to the third stage, namely, the "ML Stage," which continues today.

From the 1990s, the third stage of AI started with the objective of solving the bottlenecks in knowledge acquisition by the use of the emerging ML discipline. During this stage, for the first time, IBM's Deep Blue computer defeated a chess master Garry Kasparov, sparking a wave of AI research. After entering the 21st century, the advances of deep learning (DL) and big data technology brought about a new peak in AI research, called a new generation of AI (AI 2.0). Currently, the hot topic of AI mainly focuses on the theory and method of how to use computers to analyze data, which in turn, can be considered as the theory and methods for intelligent data analysis. Especially in 2006, AI began to enter the cognitive intelligence era emphasizing big data accumulation, theoretical algorithm innovation, computing power improvement, and self-learning. Moreover, AI has made breakthroughs in many application fields, ushering in another AI prosperity. In 2010 and 2011, scholars who had made outstanding contributions to ML were awarded the highest award in computer science, namely, the A.M. Turing Award. In the history of Turing Awards, this extremely rare situation directly reflects the importance of ML in the field of AI. Another milestone of AI 2.0 after 60 years happened in 2016, when Google's AlphaGo defeated Sedol Lee by 4:1, and almost immediately also defeated Jie Ke, who was ranked first in the world, by a score of 3:0 in May 2017, marking the end of the "human machine war" [1].

1.1.2 The artificial intelligence in energy systems

The traditional energy system is undergoing a major transformation with the rapid development of renewable energy technologies, which consequently facilitates the huge diversity of energy resources, wide distribution of energy supplies, bidirectional flows of electricity, increased deployment of energy storage, massive streams of data collected by the internet of things (IoT), and the evolving role of utilities and consumers. However, due to the small number of resources that can be controlled automatically, many system-operational decisions are still taken and enacted manually, or with a basic level of automation. Accordingly, the developments of well-integrated AI will allow more automated control resources to respond to the requirements of numerous stakeholders (such as, consumers, generators, transmission and distribution power grids, consumers). AI is applied in almost every type of renewable energy (wind, solar, geothermal, hydro, ocean, biological, hydrogen, and hybrid) for the generation, transmission, distribution, optimization, estimation, management, policy, and so on. This advanced level of control enables optimization of the system with more distributed power plants while maximizing system flexibility and reducing the cost of system operation with high proportions of renewable energies. As a result, the role of AI and big data is evolving from a facilitation and optimization tool to a necessity for intelligent and rapid decision-making [2].

As mentioned earlier, AI and other digital technologies can support the renewable energy sector in a number of ways. Currently, most of the AI-enabled advancements are in advanced weather, renewable energy generation forecasting, and predictive maintenance. In the future, AI as well as big data will further enhance decision-making and planning, condition monitoring and inspecting, supply chain optimization and the general increase of the efficiency of energy systems. The global energy system is currently undergoing a dramatic transformation, and it will continue to become more decentralized, digitized, and decarbonized in the coming decades. In order to meet the commitments made under the 2015 Paris Agreement—limiting the global temperature rise to well below 2°C—this transition must accelerate. Energy systems have become increasingly digitized in recent years, and it is clear that further digitalization will be a key feature of the energy transition and a significant driver of the industry's progress toward ambitious climate goals. Advancing technology, falling costs, and ubiquitous connectivity are opening the door to new models of energy production and consumption. Digitalization holds the potential to build new architectures of interconnected energy systems, including breaking down traditional boundaries between demand and supply [3]. A simple demonstration of the application framework of AI technology in the energy system is shown in Fig. 1−2.

1.1.2.1 Digitalization is an enabler to swift and coordinate the energy transition

This is where digitalization comes in: it is the critical factor to linking the different sectors into the most reliable, affordable, and cleanest energy system. Optimizing each sector individually will eliminate options that generate flexibility and narrow the scope of the system–wide transformation process, which accordingly, would maximize the benefits of digital technology for the entire energy system, as well as for the wider economy, the environment, and society. Digital technologies already automate complex processes, orchestrate disparate systems, and facilitate information sharing across the energy sector, and software already plays a significant role in managing our energy systems [4]. With the explosion in the availability of data, and as performance continues to improve, digital technologies will play an increasingly central role in driving a swift and

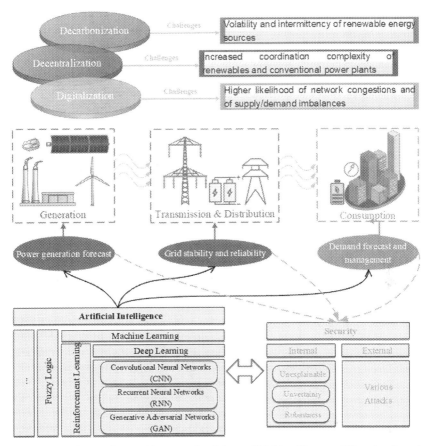

Figure 1–2 The application framework of the artificial intelligence (AI) technology in energy systems.

cost-efficient energy transition. These technologies will facilitate performance improvements and cost savings through a combination of automation, optimization, and the enabling of new business and operational models both within and beyond the traditional value chain of generation, transmission, distribution, trade, and consumption.

1.1.2.2 Decarbonizing the power sector is the starting point for full-system decarbonization

To achieve deep decarbonization, it will be necessary to shift swiftly to an energy system with very little or zero carbon dioxide emissions. The efforts to decarbonize our energy system are leading to an increasingly integrated and electrified energy system with much more interaction between the power, transport, industry, and building sectors, and a system that will consist of interdependent energy and telecommunication networks. To accelerate the shift toward a widespread, affordable, low-carbon energy supply, there is a need for greater optimization of every aspect of this energy system, as well as greater coordination and cooperation between each component. This requires a better understanding of, and better mechanisms to monitor and control, the ways in which power grids, buildings, industrial facilities, transport networks, and other energy-intensive sectors integrate and interact with one another.

1.1.2.3 The future power system looks highly decentralized

The move toward greater proportions of renewable energy generation has two main practical consequences: the future power system will host more power from intermittent power generators (since solar panels only produce when the sun is shining, and wind turbines when the wind is blowing), and it will be more decentralized. AI is a powerful tool for managing the complexities of the global energy transition and enabling greater system efficiency, reducing costs, and increasing the speed of transition. AI technology has the potential to promptly accelerate the energy transition, especially in the power sector. In this section, we identify some emerging AI applications for the energy transition in four focus areas: renewable energy generation, power grid operation and optimization, energy demand forecasting, and materials discovery and innovation [5]. AI applications can be further categorized based on the data inputs they use, such as audio, speech, images, videos, data gained from sensors, and data collected manually or robotically. As shown in Fig. 1−3, the majority of AI applications fall in the following categories of data: market, commodity and weather data. Based on the data collected from various sources, such as electricity consumption data, electricity price data, and weather data, AI is used to identify patterns or provide probabilistic predictions of future outcomes based on patterns identified in data.

The data used are usually time series data, which is a series of data points collected over a certain time interval and arranged in chronological order. By making use of image and video data to allow AI to recognize objects or conditions in images (e.g., using satellite images to determine cloud cover patterns to predict the output of a solar power plant). Meanwhile, input data are used from "smart" devices that combine sensors with communications and networking capabilities to enable real-time digital connectivity

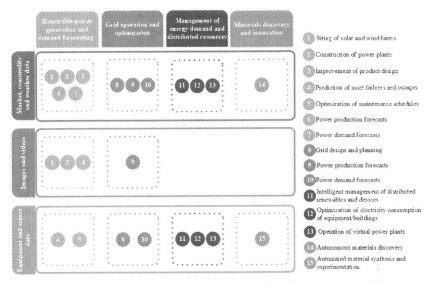

Figure 1–3 Applications of artificial intelligence (AI) for the energy transition.

and coordination of physical assets. These sensing and device-level control systems are a prerequisite for the intelligent coordination and automation of the energy system using AI. Fig. 1–3 shows the most promising applications of AI in energy conversion, categorized by the type of input used, namely, a specific application can use multiple input types [6].

1.1.3 The security of artificial intelligence–driven energy

The future energy system will become increasingly reliant on digital technologies, with millions of new connections forming a network of devices, organizations, and platforms. On the one hand, digitization promotes many positive changes of the traditional energy system, but on the other hand, as a double-edged sword, it also opens the door to increased energy security risks, given that increased connectivity across energy production, transmission, and distribution provides new opportunities for cyber hackers to exploit. Meanwhile, with digital technologies changing so rapidly, there are many unknowns about how technology, behavior, and policy will evolve over time and how these dynamics will affect future energy systems. Moreover, issues related to the security of data and data ownership might hinder the implementation of digital technologies. The risk of vulnerable data getting into the wrong hands and cyberattacks threating production are a concern for many. Therefore robust cyber security and data privacy practices will be crucial to the system stability, and to give consumers the confidence to engage with connected low carbon technologies. Furthermore, cyber security should be embedded into new systems by design, to reduce the need for costly retrofitting in future [7].

1.1.3.1 Cyber threats to energy security

There are a number of challenges that could slow the implementation of digital technologies and prevent the full potential of Industry 4.0 from being realized. One of the main challenges is the security of data, as cyber-attacks are becoming more common, threatening the continuity of operations and the theft of valuable information and availability of energy. Some cyber-attacks target operational technology (OT): computers, software and networks for the control, monitoring, management, and protection of energy delivery systems. Other attacks may only target the IT business systems that do not control the physical process of energy transfer, but lead to administrative interruptions.

The development of IoT and changes in digital technologies are increasing the potential "cyber-attack surface" in the energy system. The expansion of the IoT, coupled with the diversification and decentralization of energy technologies, is connecting millions of new small-scale prosumers, namely, the new energy producers and suppliers. However, the introduction of advanced and intelligent technologies into the energy sector presents both opportunities and challenges. The increasing number of connected devices has provided a vast surface area for attacks that exploit IoT devices with weak security. For instance, if there is one suspicious device at the edge of a network, it could be a weak point for the entire system. Moreover, the growth of smart energy has also resulted in an exponential increase in network intelligence across the energy grid and consumer premises. Consequently, a massively expanded "attack surface" constitutes the operational basis of the current energy ecosystem. At the same time, the energy system is also fundamentally interconnected with almost every other critical infrastructure network, not just the electricity grid, but the highly interconnected and interdependent natural gas, water, communications and fuel distribution systems, etc. Hence, the cybersecurity threats to the energy sector could affect every aspect of our modern society.

2012
Havex Malware

Havex is malware that has been targeting the energy sector since August 2012, and it has been developed to implement industrial espionage against a number of companies in Europe. Within the energy sector, this malware particularly targeted energy grid operators, major electricity generation companies, petroleum pipeline operators, and industrial equipment providers. Havex was distributed via spear-phishing attacks and spam emails. The victims identified were mainly from the US and Europe.

2016
CrashOverride

CrashOverride is malware that has affected a single transmission level substation in Ukraine's power grid on December 17, 2016. The incident resulted in parts of the capital of Ukraine, Kiev, to plunge into darkness, causing a blackout. The outage lasted for an hour. This was the first-known case of malware created to particularly hit electrical grid systems. Once it infects Windows machines, it automatically maps out control systems and records network logs to send to its operators.

2018
Operation Sharpshooter

Operation Sharpshooter was an espionage campaign disclosed in December 2018. This was a novel implant framework to hack global critical infrastructure players including nuclear and energy companies based in Germany, Turkey, the US, and the UK. The espionage campaign started when a splay of malicious files were delivered to targets via Dropbox. Once downloaded, these files placed embedded shellcode into the memory of Microsoft Word.

2015
GreyEnergy

GreyEnergy is an Advanced Persistent Threat (APT), which targeted industrial networks in Ukraine and also Eastern European countries for the last several years. Security researchers have linked GreyEnergy with Black Energy, an APT targeting Ukraine and leaving 230,000 people without electricity in December 2015. This malware majorly utilized phishing emails as its initial infection method.

2017
TRITON/TRISIS malware

TRITON/TRISIS, a dangerous malware initially discovered in mid-November 2017, was utilized to attack framework to control industrial safety systems at a critical infrastructure facility and accidentally led to a process shutdown. The malware targeted the Safety Instrumented Systems (SIS), specifically Schneider Electric's SIS, the Triconex Emergency Shut Down (ESD) system. FireEye, a cyber security solutions provider, stated that the intrusion tools the malicious actors used were developed by humans, that they had perceivable human strategies and preferences

Figure 1−4 Notable cybersecurity attacks on energy sector.

1.1.3.2 Notable cybersecurity attacks on the energy sector

As depicted in Fig. 1−4, cyber-attacks on the energy sector are becoming increasingly targeted and sophisticated, their impact can be devastating for energy companies as well as for the public. Yet, the methods of cyber-attacks constantly change, which means that a defense system that operates perfectly today may not be efficient tomorrow. A number of adversaries, each with their own motivation, strove to compromise organizations that operate critical infrastructure. Below are some of the most notable cybersecurity attacks that caused an intensive impact on the energy sector [8].

1.1.3.3 The countermeasures to cyber-attacks on energy sector

Complete protection against cyber-attacks is impossible, but their impact is limited if countries and companies are well prepared and inherently resilient. This is especially significant for critical infrastructure: that is, physical and institutional assets that are vital to the functioning of the economy, such as large-scale energy systems. Developing system–wide countermeasure depends on all personnel and stakeholders being aware of the risks first. The success of any attack depends not only on the capabilities of the attacker, but also on the vulnerability of the target and its readiness to respond. Digital energy security should be built around five key concepts:

1. Issuing proactive policies on cyber security at regional and national levels: overall, the warning signs for the industries and societies are apparent, and should lead to an effort to put stronger, more modern, and reliable protections in place, to benefit all stakeholders.

2. Using AI to increase cyber security: due to the growing volume of digital data, the use of AI in defending against attackers hiding in the abundance of digital signals is also pivotal. AI solutions can spot abnormalities that will, in most cases, not be apparent to the human eye and enable more proactive mitigation of the impact of or even fight off the attacks themselves. AI can also ensure a nation, system, or institution can adapt to changing contexts, to withstand shocks, and to quickly recover or adapt to a desired level of stability, while preserving the continuity of critical infrastructure operations [9].

3. Promoting cyber hygiene: the basic set of precautions and monitoring that all information and communication technology (ICT) users should undertake. This includes first and foremost building awareness. Other key elements are secure configuration of equipment and networks, keeping software up to date, avoiding giving staff and users unnecessary system privileges or data access rights, and training to establish a security-conscious culture throughout an organization. For example, staff may need to be trained on the secure use of their personal computers at home and mobile devices.

4. Considering security in the initial design: the incorporation of security objectives and standards as a core part of the research and design process; security should not be a later add-on after a product has been built or supplied to users. Security by design can be an efficient way to reduce overall risk.

5. Strengthening end user responsible practice: the hyperscale cloud computing providers certainly benefit from no or less shortage of skilled resources to implement and operate at the underlying service platform at the highest possible level of security and compliance. The end user (data controller) is nevertheless also responsible for implementing the

"last mile" of security in configuration and everyday usage by their employees. The tasks for the customer in protecting the integrity, confidentiality, and availability, as well as establishing resiliency through technical architecture and organizational implementation of security measures are easily overlooked, but should in no way be considered trivial.

In general, regulatory standards will struggle to keep up with rapid technological changes and new vulnerabilities. Governments and energy companies, therefore, need to be both proactive and adaptive. In some cases, innovative solutions might be found when energy companies work together.

1.2 Carbon neutrality from energy ecosystem

1.2.1 Natural and energy ecosystem

Carbon on the earth is mainly stored in rocks, soil, atmosphere, hydrosphere, and biosphere, and frequently exchanged and recycled in the atmosphere, hydrosphere, and biosphere. The carbon cycle related to human activities is mainly contained in the biological cycle of carbon. Terrestrial and aquatic autotrophs (such as plants and algae) absorb CO_2 in the atmosphere through photosynthesis, then convert it into carbohydrate and fix it in plants. The corresponding carbohydrate will further partially transfer to animals through the food chain. At the same time, animals and plants will release some carbon into the atmosphere in the form of CO_2 through respiration. After animals and plants die or form wastes (such as fallen leaves and feces), carbon in organic matter is converted into CO_2 through microbial decomposition and discharged into the atmosphere. In addition, a small part of animal and plant residues are buried by sediments, converted into fossil fuels, and finally burned by humans to obtain energy, emitting a large amount of CO_2. Besides, the mass death of plants may also be caused by human factors, such as deforestation. However, forests with less alkali cannot absorb CO_2 anymore, and will become carbon sources, which is CO_2 generated by land use change. This typical carbon cycle model and the role of human activities in it are shown in Fig. 1−5.

It can be seen from Fig. 1−5 that in the biological cycle of carbon, when there is no human intervention, CO_2 is released to the atmosphere

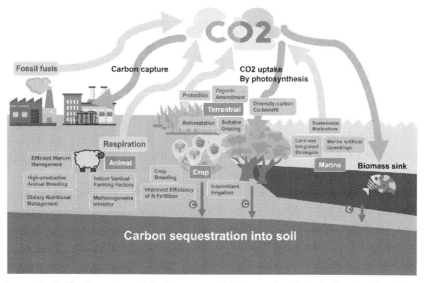

Figure 1—5 The impact model of human activities on the biological cycle of carbon.

mainly through the decomposition of biomass by microorganisms and the respiration of animals and plants, which is the main carbon source. The absorption of CO_2 in the atmosphere is mainly achieved through the light cooperation between terrestrial and aquatic plants (as well as algae, some bacteria, etc.), which is the main carbon sink. In the natural state, the release and absorption of CO_2 have reached a dynamic balance. A small increase of CO_2 in the atmosphere will promote the growth of plants and photosynthesis of plants, thus reaching a new balance. However, the burning of fossil energy and the destruction of forests and marine ecosystems by humans have broken this balance, resulting in the continuous and rapid growth of the content of CO_2 in the atmosphere. To reverse this situation, the basic idea is to reduce CO_2 and emissions, and increase CO_2 and absorption capacity. Finally, the absorption rate of CO_2 will reach or exceed the emission rate, so as to achieve carbon neutralization. Carbon peaking is an important stage in the process of carbon neutralization, which means that through efforts, the net emissions of CO_2 will no longer increase, and continue to work hard to achieve the reduction of CO_2 and net emissions, and finally achieve carbon neutralization.

Carbon emission reduction is the key to the sustainability of modern society [10]. With the intensification of global industrialization and the overexploitation of nonrenewable energy, a large amount of greenhouse gases are released, leading to the rise of global temperature and a series of environmental degradation problems [11]. From around 1850 in preindustrial times to 2022, the global average atmospheric CO_2 concentration increased substantially from 285 ppm to 419 ppm [12]. In order to cope with the rise of global greenhouse gas concentration and temperature, 197 member states of the United Nations Framework Convention on Climate Change (UNFCCC) unanimously agreed to adopt the Paris Agreement at the Paris Climate Change Conference (PCCC) on December 12, 2015, which has formulated a global action plan for coping with climate change after 2020 [13]. According to the Paris Agreement, countries agreed to limit the global temperature increase to less than $2°C$ and strive to limit the global temperature increase to less than $1.5°C$ [14]. As of February 2021, 124 countries worldwide have declared their intention to become carbon neutral and achieve net zero carbon emissions by 2050 or 2060 [11]. In order to achieve the goals set out in the Paris Agreement and support sustainable development, it is necessary not only to reduce CO_2 emissions, but also to remove CO_2 from the atmosphere through various social, economic, environmental, and technological measures to achieve net zero or negative carbon emissions.

Carbon neutrality is a net-zero carbon emission status that can be achieved through carbon offsetting or removal initiatives that balance the total amount of carbon dioxide or greenhouse gas emissions directly or indirectly produced by a country, company, product, activity, or individual over a given period of time. In addition, in order to achieve carbon neutrality, the Intergovernmental Panel on Climate Change (IPCC) in its special report on global warming of $1.5°C$ also stressed the need to reduce and phase out fossil fuels, use more renewable energy, improve energy efficiency, and stressed the importance of implementing these measures in cities to achieve carbon neutrality. In addition, carbon removal or sequestration in terrestrial and marine ecosystems must be promoted in order to achieve net zero carbon emissions and sustainable development [15]. Different regions, countries, and cities have developed strategies to improve carbon removal or sequestration and achieve carbon neutrality [16−18], but achieving net zero carbon emissions is a challenge.

The countermeasures to reduce carbon emissions and achieve carbon neutrality can be divided into four main ways: carbon substitution, carbon emission reduction, carbon sink, and carbon cycle. Carbon substitution refers to the use of hydropower, photovoltaic, wind power, or other "green electricity" to replace thermal power. Heat substitution refers to the use of light, heat, geothermal, and other fossil fuels to replace heat supply. Hydrogen substitution refers to the use of "green hydrogen" to replace the hydrogen generated from fossil energy. Carbon emission reduction includes energy conservation and energy efficiency improvement. In the construction industry, it mainly focuses on the improvement of energy efficiency for electrical appliances and equipment, the implementation of solar PV on houses and the development of new low-carbon emission materials for cement and steel. In the transportation industry, the methods for achieving carbon emission reduction are more efficient power systems and lighter materials, which reduce carbon emissions from the source. Carbon sink refers to the collection of carbon dioxide generated by large thermal power plants, steel plants, chemical plants, etc., and then transportation to appropriate places, achieving long-term isolation from the atmosphere. The carbon cycle includes artificial carbon conversion and forest carbon sink. Artificial carbon conversion refers to the conversion of carbon dioxide into useful chemicals or fuels by chemical or biological means, including the synthesis of methanol from carbon dioxide, and the preparation of CO or light hydrocarbon products (C1–C3) by electrocatalytic reduction of carbon dioxide. Forest carbon sink refers to the absorption and fixation of carbon dioxide in the atmosphere by plants through photosynthesis in vegetation and soil to reduce the concentration of carbon dioxide in the atmosphere.

From the perspective of world energy development, the history of human energy utilization has experienced two transformations from firewood to coal, coal to oil and gas, and is undergoing the third transformation from fossil energy to renewable energy. Due to the lack of large-scale energy storage, the full utilization of renewable energy is still facing major challenges [19]. Here, we have introduced the energy ecosystem to combine the energy cycle with the carbon material cycle, so as to control the energy balance and carbon emissions from a systematic and overall perspective. We define energy ecosystem as a developing life system, including industrial energy, transportation energy, building energy, and related carbon negative processing cycle chemical materials. The similarity with natural ecosystems is that they all have energy and material cycles to create a flow path for sustainable development [20]. The difference is that natural ecosystems are self-organized by nature, while energy ecosystems are manufactured by artificial intelligence. The goal of the energy ecosystem is to simulate the behavior of the natural ecosystem, so that it can achieve the maximum use of resources at the minimum cost.

1.2.2 Energy ecosystem roadmap

The first aspect of energy ecosystem is energy consumption, which is mainly distributed in residential, commercial, industrial, and transportation applications [21]. All residential and commercial applications can be further redefined as architectural applications. Therefore in 2022 the global industrial energy consumption will be about 30%, the traffic energy consumption will be about 25%, and the building energy consumption will be about 45%. The different development stages of the three main energy applications are summarized in Table 1−1.

As can be seen from Table 1−1, intelligent design and operation have become the goal of all applications, which is consistent with the recent interest in the development of artificial intelligence technology. Similarly, we predict that the smart energy industry is on the way, and we define it as an energy ecosystem, including relevant activated carbon cycle control different from other similar concepts and definitions. The roadmap for energy ecosystem development also includes four steps, as shown in Fig. 1−6. In these four steps, the first two steps have been determined by the industry, and the energy society is building the third step. There are some major challenges, as shown below:

- Decentralization: A real distributed infrastructure need be formed to replace the existing centralized one.
- Real time: A real time processing capability need be formed from the dispersed architecture.
- Energy buffering: Energy storage from massive energy to high power are both needed due to flexibility.
- Resource utilization: Large data for processing with communication bandwidth and data storage limitation [22].
- Self-learning: Self-organization and self-immunity require artificial intelligence in energy system [20].

Table 1−1 Development stages of major energy applications.

Category	Digitation	Network	IoT	Smart
Industrial	Digital equipment	Industrial automation	Industrial IoT	Smart Industry 4.0
Transportation	Digital vehicle	Vehicle automation	Internet of vehicles	Self-drive
Building	Digital system	Building automation	Smart home	Smart city

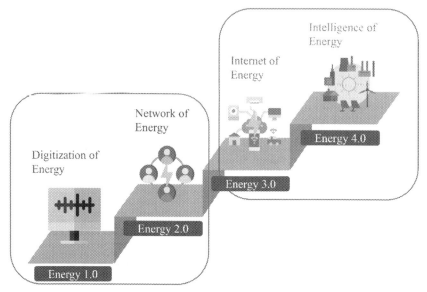

Figure 1−6 System roadmap of the energy ecosystem.

Therefore this paper proposes an intelligence energy ecosystem as the systematic and holistic solution to overcome the issues above.

1.2.3 Carbon neutrality in energy ecosystem

To better understand the concept of energy ecosystem, Fig. 1−7 shows a fundamental concept of this solution. The whole diagram is classified into many conversion stages defined as P2P (power to power), P2H (power to hydrogen), H2P (hydrogen to power), G2P (gas to power), C2P (coal to power), H2G (hydrogen to gas), P2M (power to mobiles), and P2C (power to chemicals) and all together they can form the carbon cycle within energy ecosystem as shown in Fig. 1−8. The CO_2 produced by combustion for transportation or heating supply or generation was captured and then was electrolyzed as fuel cell. It also can be used for synthesis and crude chemicals. Then after refining and purification, it can be used as distribution energy.

The carbon cycle within the energy ecosystem integrates carbon capture and utilization technology, which is an effective way to achieve carbon emission reduction. CO_2 capture, use and storage (CCUS) technologies involve three distinct processes: separation of CO_2 from emission sources, CO_2 conversion and use, transport, and underground

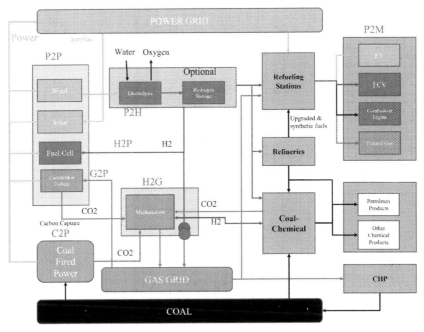

Figure 1−7 Solution concept diagram of the energy ecosystem.

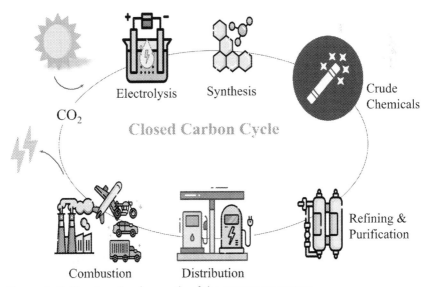

Figure 1−8 The closed carbon cycle of the energy ecosystem.

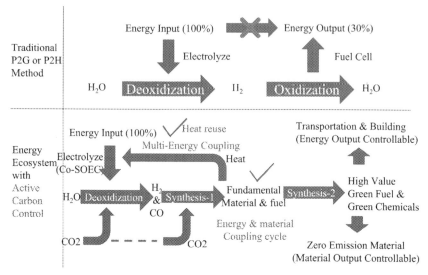

Figure 1–9 Coupling between energy flow and material flow.

storage in long-term isolation from the atmosphere [11]. Ecosystems work by coupling energy flow and carbon material flow in a closed loop, which makes them different from traditional P2G or P2H concepts, as shown in Fig. 1–9. For traditional P2G or P2H methods, the efficiency is only 30%. In an energy ecosystem with active carbon control, heat is reused during the synthesis process, and energy flow and carbon matter flow are considered together to achieve a coupled energy and matter cycle. By actively controlling the demand for energy or chemicals, the ecosystem can maximize resource utilization with minimal cost from the perspective of the whole system [23,24].

1.3 From industrial revolution 4.0 to 5.0

1.3.1 From Industry 1.0 to Industry 4.0

The fourth industrial revolution brings unprecedented opportunities to the world, but also faces huge challenges. Different from the previous three industrial revolutions, which occurred in late 1800s, early 2000s, and mid-2000s, respectively, with the industry (transportation), energy, and information technology as breakthrough points, the fourth industrial revolution as shown in Fig. 1−10 is in the context of the development of the digital economy, through the wide application of artificial intelligence (AI) technology, forming a new service-centric industry form. Indeed, Industry 4.0 is about the digitization of resources, digitalization of processes and leveraging of the new environment. By leveraging the digital space, Industry 4.0 is reinventing industry. For example, Fig. 1−11 shows the business models development under the digital economy. The current trend of in-depth integration of the digital economy and the real economy has brought new opportunities and challenges to the energy industry. Combining the innovative application of emerging technologies will surely reshape productivity and production relations, and inject new momentum into the transformation and innovation of the energy industry through digital transformation. The intelligence of the energy industry is undergoing a development path from mechanism-driven to data-driven, then to autonomous intelligence, showing a trend of high integration with the information industry and the transportation industry. Therefore the nature of the energy revolution under the digital economy is actually a combination of AI and energy technology. In the interaction with the information revolution and the transportation (mobility) revolution, it continues the development process model of the digital economy from e-commerce mode to sharing mode, then to "X as a Service (XaaS)" mode [25].

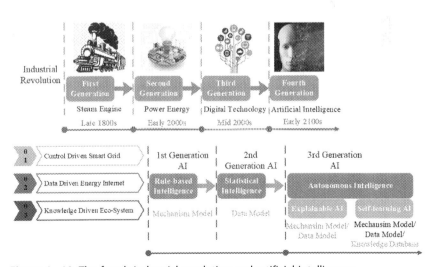

Figure 1−10 The fourth industrial revolution and artificial intelligence.

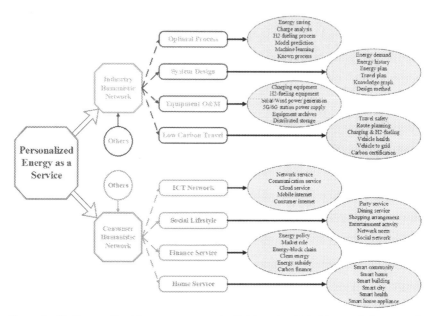

Figure 1–11 Personalized "energy as a service" driven by digital humanistic network.

However, the Industry 4.0 is addressing just part of the whole. In fact, the changes brought about by Industry 4.0 involve the whole society, and this broader transformation should be given a new name, namely Industry 5.0. Rather than just focusing, as we often do, on the extent to which the "digital transformation" may affect jobs, it is necessary to look more at the whole transformation of societies with the need to reinvest in human capital through a different education program, involving a re-training of current workers to use them in different-new-sectors [26].

Industry 5.0 captures this point, where the relationship is no longer that "using (smarter) machines" can eventually replace humans, but that machines are harnessed through collaboration between machines and humans. It is a two-way relationship, humans learn from machines and machines learn from humans. This requires a move to cognitive machines, one of the novelties in Industry 5.0. Another crucial point to be addressed by Industry 5.0 is to go beyond the industrial environment, taking into account the whole system, that is considering the impact that industry has/may have on the whole. In the present and in the future, sustainability is both economical and environmental. In addition, industry is not only required to have a zero impact on the environment, but must also contribute to the improvement of the field.

1.3.2 The Industry 5.0

While Industry 4.0 is characterized, and fostered at the same time, by technology adoption such as internet of things (IoT), smart robotics, AI, 3D printing, connectivity 4G/5G, and digital twins (DTs), the shift toward Industry 5.0 has three distinct features, namely, human centricity, reliability, and sustainability. These are not "technologies," although there are numerous technologies supporting them, and of course, investment in technology evolution is also needed. Meanwhile, the aforementioned technologies remain crucial given that Industry 3.0 cannot jump to Industry 5.0 directly, it is necessary to implement Industry 4.0 to get to 5.0. From another point of view, Industry 4.0 is driven by "internal" pressures to become more efficient and flexible, to reduce capital expenditure by using more flexible production tools such as smart robots, and at least keep income stable as prices slide due to digitization transformation. Nevertheless, Industry 5.0 is driven by "external" pressures, aided by technological evolution, to become an integral part of global sustainability efforts, which include reduced raw material consumption, CO_2 reduction, recycling and reuse, to take care of an evolution-focused workforce a human–centric but not business-centric approach, and become more resilient and sustainable. In other words, the relationship among human and machines in Industry 5.0 will be much different from that in Industry 4.0. The machines will need to evolve from being tools, or alternative to human labor, to becoming partners and this involve increasing their cognitive capabilities. The meaning of education and knowledge is evolving, and it is important to recognize the influence of technology in shaping these concepts. In addition to traditional cultural education, we must also acknowledge the increasing role of machines in accessing and interpreting information. As a result, the ability to utilize and apply knowledge (i.e., knowing how) may become more crucial than simply possessing factual information (i.e., knowing what) [27].

Among the human-centricity, reliability, and sustainability in Industry 5.0, the humanity plays a most significant role and also the most distinct role from Industry 4.0, which focuses more on the technology itself. The general definition of humanity refers to the advancement and core parts of human culture, that is, advanced values and custom. Culture is the symbols, values, and customs shared by human beings or a nation or a group of people. Symbol is the foundation of culture, while value is the core of culture. Custom, including living custom, moral custom, and legal custom, is the main content of culture. The rapid development of mobile internet and digital social network has fundamentally changed people's lifestyles of shopping and socializing, forming a new type of digital consumer humanistic network.

This chapter will propose a new concept defined as industrial humanity. Industrial humanities are the advanced and core part of the industrial ecology which includes the advanced business models and industrial standards in the industry. Industrial humanities, as shared by industrial participants, include industrial technology methods, business models and standards. Industrial technology methods are the foundation of industrial humanity. Industrial business models are the core of industrial humanity. Industrial standards, including technical standards, safety requirements, legal rules, and domain knowledge, are the main content of industrial humanity. Driven by the internet-of-everything, the industrial humanistic network and the consumer humanistic network form a new digital humanistic network, which reflects the new production relationship between people, people-and-things, and things-and-things under the digital economy. How to understand the interaction and fusion significance of this new type of production relationship for new digital productivity is the key of discussion in this chapter.

There is a sustainability reason behind the shift from Industry 4.0 to Industry 5.0. We need to put humans at the center of the equation again. Industry 4.0 is the reinforcement of the unsustainable industrial production paradigm (machine-centric), with the help of IoT and AI, we can move into a results-based economy (human-centric) sustainable postindustrial production paradigm. Moving from ownership of things to performance of things leads to more profit for companies, more health for people, and no harm to the environment [28].

Sustainability addresses a variety of issues, from the focus on the well-being of people and the development of the economy, to attention to the production, consumption, and pollution of resources. At the same time, economic sustainability is not just a company or industry issue, it is a systemwide global issue that each country should take into account. For instance, due to the high cost of generating power from sustainable power plants, a company will not be able to generate revenue since the price of its product is noncompetitive in the market. Correspondingly, an industrial production paradigm based on renewable sources of power may be economically unsustainable at a company level. However, by adopting tax-exemption or giving subsidies, it could become sustainable if government action sustains the utilization of renewable energy by lowering its price or by imposing on all businesses in a given market mandates to use renewables, thus leveling the market rules. As mentioned earlier, the market itself can steer the industry toward environmental sustainability by choosing not to buy from companies that are not concerned with the environment, and willing to pay more for environmentally friendly products. However, this market influence is not as strong as "imposed" government influence. There are certainly plenty of environmentally sensitive customers willing to pay more to protect the environment, but the majority are most likely to buy at a lower price.

It should be clear that a shift to Industry 5.0 with pursuing societal and environmental objectives will result in a huge cost to industry. Industry is willing to pay for this, as long as the added cost does not make its portfolio economically unsustainable. Therefore there is a need to think outside the industry "box" and take a system-wide perspective. Moreover, the shift requires effort and energy that will not spontaneously come from industry.

1.3.3 Technology enablers of Industry 5.0

Initially, technology is a tool, while some technologies have shaped the evolution and actually prompted evolution. The very first technologies such as stone cutting and shaping, bronze smelting, and iron producing were used to give names to entire epochs, namely, the stone, bronze, and iron ages. More recently, we have used techniques to describe our industrial revolution, such as the use of steam engines, electricity, computers, and AI to represent Industry 1.0 to Industry 4.0. This does not seem to be the case with Industry 5.0, where all technologies used are also part of Industry 4.0, possibly to varying degrees [29]. For example, edge computing, AI, robots, high-speed communication, DTs, industrial blockchain, IoT, and big data analysis as illustrated in Fig. 1−12.

Cloud and edge computing: Cloud computing enables the handling of large amounts of data that are needed to automate production processes, and to achieve extraordinarily short reaction times. It will become ubiquitous and irreplaceable. Edge computing enables that products both benefit from edge computing and become part of the edge, thus increasing edge processing capabilities, this further increases resilience and decreases the need for additional resources. It will promote solutions such as infrastructure as a service (IaaS), platform as a service (PaaS), software as a service (SaaS).

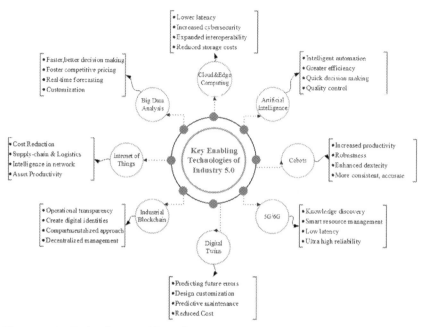

Figure 1−12 Technology enablers of Industry 5.0.

Artificial intelligence: AI is already the bread and butter of Industry 4.0, but it is evolving so fast that the AI that will permeate Industry 5.0 will be much larger in terms of performance, thus making the design, production, and use of products more efficient. Most importantly, this ever-expanding AI will go hand in hand with the worker on the shop floor and throughout the supply/delivery chain, finally embedded in the product and around the user, enhancing his intelligence. This will become one of the pillars of Industry 5.0, a powerful human-centric tool. From a company's point of view, AI supports intelligent automation and all steps in the life cycle to a greater extent than Industry 4.0. The real difference is whether it is possible for AI to conform to an externally agreed framework for protecting human labor, so that humans remain at the center of the process.

Digital twins: DTs are fundamental tools in Industry 4.0 and have a tendency to become increasingly common by flanking products in Industry 5.0. They will link usage to the lifecycle, provide feedback, and enable continuous improvement of the design. DT will contribute to resiliency throughout the lifecycle. Furthermore, the interaction of equipment DTs on the shop floor and supply/delivery chains with the workers' personal DTs can be foreseen, another distinguishing element of Industry

5.0 from Industry 4.0. These interactions should empower workers and keep them in control.

Cobots: More and more collaborative robots (Cobots) are now appearing on the shop floor. Their numbers are sure to increase in the coming years and each one is expected to interact in the metaverse through their DT. They provide flexibility to support the resilience goals of Industry 5.0. Advances on Industry 4.0 are primarily related to the possibility of each Cobot gaining a "full picture," that is, understanding the environment inside and outside the workshop, including supply chains, warehouses, and product usage, and also the cooperations with humans. Advances in flexibility will also support environment-conscious designs by considering end-of-life, refurbishment, and recycling.

5G and 6G: In Industry 4.0, cyberspace is more collaborative through the entire value chain and ecosystem, which requires high-performance communication infrastructure. With the advent of Industry 5.0, manufacturers, users, and third parties will all place increasing emphasis on controlling the "operation" of products. This will require a pervasive communication infrastructure, which can also be created by converting products into communication nodes. This is part of the 5G architecture that will be an essential feature of 6G. Developments in communications infrastructure will make it possible for products to operate with greater efficiency, thereby reducing power consumption. In addition, product functionality will also be delivered through the cloud, potentially extending the product's life cycle and service life, reducing resource requirements and waste. It should be noted that this goes against the "consume and throw away" culture of the past 40 years. This, in turn, will require profound changes in the industry and financial markets, led by ever-increasing revenue (i.e., selling more). In addition, the 6G-enabled network architecture will support seamless communication among DTs and personal DTs, with all the local processing and bandwidth required by the metaverse, accessible through augmented and virtual reality, thus supporting human-centric goals.

Industrial blockchain: The blockchain is already wide-ranging in the financial and banking markets, and it will rapidly move to logistics and manufacturing in the coming years. Moreover, it will be continuously refined to take more consideration of sustainability and will surely be a widely adopted tool in Industry 5.0. Compared with the traditional process which is to track every single component ending up in the final product, the industrial blockchain provides a more convenient and reliable way of tracking, taking care of verifying the different steps in the manufacturing process, starting from the supply chain up to the delivery chain. Specifically, as metaverse will become the way of production and life, the need for trust and control will grow exponentially, and block-chain is the key technology that can meet these growing needs.

Big data analytics: Industry 4.0 is using and creating massive amounts of data. This has led to the increasing adoption of data to repre-sent and analyze any aspect of the production process and product usage until its demise. All of these data are creating the soft infrastructure needed for the product lifecycle and, to an increasing extent, delivering the perceived product functionality. This goes a long way toward eco-nomic and environmental sustainability, as well as resilience. From a man-ufacturer's perspective, it is a no-brainer: data facilitates customization, supports all decisions throughout the lifecycle, and a good set of redesigns at lower cost and greater flexibility.

Internet of Things: IoT is a ubiquitous fabric in Industry 4.0. It will extend further beyond company boundaries, supporting product designs that leverage the environment to reduce material requirements and reduce waste. The capabilities of the product can be choreographed through the cloud, especially the edge cloud, and can piggy-back on the capabilities provided by other products/infrastructures available in the environment. To support this vision, novel communication paradigms, such as those enabled by 6G, must be developed.

By moving to cyberspace and digital space, efficiency can be improved and the impact on the environment can be reduced to a certain extent. This is underscored by the name used by several countries to denote the fourth gen-eration of the industrial revolution: digital industry. Nevertheless, industry does not stand in a void: it depends on the market and on the people working in its factories and in its value chain. At the same time, the evolution of industry, of its products and of the tools used in the production also affects the society, the technology we create is changing the way we live. This is underlined by the name associated with this shift: from industry revolution 4.0 to 5.0.

1.4 Emerging sign of supersmart Society 5.0

1.4.1 What is Society 5.0?

Before getting to the topic of Society 5.0, here is a question for you: how do you envisage our society in the future? Will it be utopia where everyone could live happily and there will not be any social issues, or will it be dystopia where technologies consume us, and the planet will no longer be suitable for humans? So, we notice that there are two key concepts: social issues and technologies. The former is what we must resolve, the latter is our possible tool. Why it is a possible tool not necessary tool? It is because nowadays for many people, their lives have not been greatly improved with the existence of modern technology. On the contrary, they suffer from great cost of living and huge pressure. One may argue that it is more of a cultural or psychological reason. However, changes in the means of production determine changes in productivity, and changes in productivity drive social change. Therefore when we realize our society has become a mass production and mass consumption type with the aid of technology. We need to rethink the role of these technologies and how to put them into better use. That is how Society 5.0 comes into view.

In 2017 Japan government released its comprehensive strategy (Cabinet Office 2017), where it presented: "Society 5.0, the vision of future society toward which the Fifth Basic Plan proposes that we should aspire, will be a human-centered society that, through the high degree of merging between cyberspace and physical space, will be able to balance economic advancement with the resolution of social problems by providing goods and services that granularly address manifold latent needs regardless of locale, age, sex, or language to ensure that all citizens can lead high-quality, lives full of comfort and vitality" [30].

As depicted in Fig. 1−13, Society 5.0 is proposed to balance various interests including individuals, companies, regions, ethnic groups, environment, and so on. With the development of internet of things (IoT), artificial intelligence (AI), and Big Data, this concept will be possible because there are so many things to consider that it is not feasible for a human to come up with a solution. More powerful AI fed by various data including personal information, traffic information, economic information, and even climate information will make precise, customized decisions that are beneficial to the long-term healthy development of human beings. There are two concepts that are of great significance for our understanding of Society 5.0 [31].

1.4.1.1 Human-centered society
When we only pay attention to the economic development of society, a series of problems such as regional development imbalance and

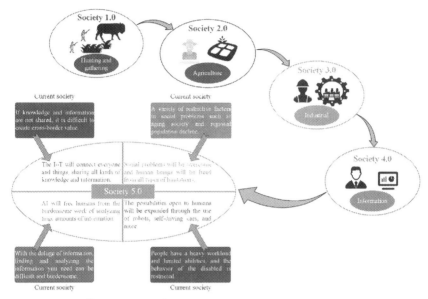

Figure 1–13 Different stages of society.

overproduction often arise. These problems are reflected in people's lives through the rapid turnover of consumer goods, promoting economic growth but resulting in low utilization and various forms of waste. When we talk about a human-centered society, we are taking a holistic interest in human beings. For instance, when using lights, we may want the room to be bright all the time, but during the day it is unnecessary to use them. Similarly, during the hot season, we may want the air conditioner to be on all the time, but we need to consider the carbon footprint of the air conditioner as well as individual needs when we are away or when some-one wants a different temperature. Compared to traditional societies, which aim to provide services to those who buy products, Society 5.0 aims to balance individual needs, social development, and problem-solving.

1.4.1.2 Merging cyberspace with physical space

In the old days, control systems combined with other technologies only allowed us to regulate the operation of a single or several systems. Nowadays, with the help of Big Data and AI, we have the capability to simulate the real world if we have the data which may need assistance from IoT. As shown in Fig. 1–14, cyberspace is the simulated world in

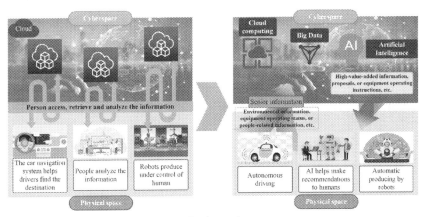

Figure 1—14 Merging cyberspace with physical space.

which we perform computations, and physical space is where we put the results of our simulations into practice. Therefore the process is as follows: collect data from physical space, analyze data and draw possible solutions in cyberspace, implement strategies in physical space, and finally test the validity of the model given in cyberspace. Society 5.0 has achieved a high degree of integration of cyberspace (virtual space) and physical space (real space). In the past information society (Society 4.0), people access cloud services (databases) in cyberspace through the internet to search, retrieve, and analyze information or data. In Society 5.0, a large amount of information from physical space sensors is accumulated in cyberspace. In cyberspace, these big data are analyzed by AI, and the analysis results are fed back to humans in physical space in various forms. In the past information society, the common practice was to collect information through the network and analyze it by humans. And in the 5.0 society, people, objects, and systems are all connected in cyberspace, and the optimal results of AI surpassing human capabilities are fed back to physical space. This process brings new value to industry and society in unprecedented ways.

1.4.2 The evolutionary direction of society

As mentioned earlier, we learned that Society 5.0 proposes that we use new technologies to merge cyberspace and physical space. In addition to this macro-level convergence, there are many other features that can convey the evolution direction of future societies.

1.4.2.1 Industry 4.0 and Society 5.0

We are familiar with the first industrial revolution in which steam engines played an important role for development of textile manufacturing, the second industrial revolution which increased electrification and facilitated the development of modern production line, the third industrial revolution where computers were widely used so that men do not need to worry about massive repetitive simple tasks. However, the fourth industrial revolution has not been understood by many, even though they may have heard about it. Basically, the fourth industrial revolution tells us the trend of automation and massive use of data in manufacturing technologies. Such technologies include cyber-physical systems (CPS), or digital twins, IoT, cloud computing, and AI [32].

Both Industry 4.0 and Society 5.0 seek help from new technologies, however, their objectives, difficulties, and implementation processes are quite different. First of all, what Industry 4.0 does is actually monitor the whole process of manufacturing and selling so as to make the factories provide more customized goods and better services while lower or maintain the production costs with the aid of technologies at the same time. Society 5.0, on the other hand, requires "balance economic advancement with the resolution of social problems by providing goods and services that granularly address manifold latent needs regardless of locale, age, sex, or language to ensure that all citizens can lead high-quality lives full of comfort and vitality" [30]. Therefore Industry 4.0 is capital-centered, while Society 5.0 is human-centered. Secondly, since the reforms envisioned by Society 5.0 will profoundly affect everyone's life, there are many nontechnical issues to define and solve. For instance, how to quantify the quality of life, how to get a full picture of latent needs and preferences, how to trade off personal interests and society interests, these problems will not appear in Industry 4.0. Finally, the design and execution of Industry 4.0 are top-down, while for Society 5.0 it is bottom-up. Simply because Society 5.0 is human-centered.

1.4.2.2 Society with intensive knowledge

Sometimes when we describe a person, we may refer to him or her as knowledgeable. That is possible because he or she knows so much. However, from the perspective of data science, data, information, and knowledge are three levels of our understanding. For example, we may easily get the basic data of a boy, his height, weight, appearance, what is he doing, etc. These things are data, once we continue to get in touch with him, we may keep noticing how often does he change his clothes, how often does he play basketball, does he like talking, etc. So from these sorted data, we obtain some information about this boy. After a long time observation, we may find he is very responsible for everything he is in charge of, he is very open-minded, and he likes to help others, therefore you want to do business with him. Using all kinds of information to come up with a decision, the integration of information and what enables you to make a decision is knowledge.

Labor-intensive society relies on massive workforce to realize large production. Capital-intensive society concentrates on tangible goods which are clustered near seaports and airports. While knowledge-intensive society thinks values exist in knowledge and the process of transferring it. How to comprehend this? When we solve a problem, the thing that hide behind our decision-making process is knowledge. We can use others similar solution, or we use their knowledge. Another word is "intensive," where are all this knowledge then? Society 5.0 requires merging between cyberspace and physical space, which include large amounts of data, humans are not able to analyze data under such huge scope, therefore AI will contribute to knowledge generation and transfer.

1.4.2.3 Zero carbon society

Climate change intensifies and energy obtained from fossil fuels shortages every year. People agree that it is urgent to transform the energy structure and reduce carbon dioxide emissions. The Chinese government promises to peak carbon dioxide by 2030 and strives to achieve carbon neutrality by 2060 [33]. Increasing the portion of renewable energy and trying to capture carbon dioxide in the air is one way to go. Even though we may encourage a low-carbon life, we still want to ensure people's well-being. So is it possible for us to lower carbon emissions and maintain our life quality? Society 5.0 says yes.

For one thing, people's demands are various. Some may favor 24°C of AC in summer, others may think 28°C is fine. The common way to satisfy most people is to set a compromise point. For instance, 26°C. However, if we give 24°C to people who want 24°C and charge them some money for they consume more energy than others, and let everyone who can satisfy with 26°C or more not need to pay. There will be more people getting used to 26°C, and people who use 28°C AC will save some electricity for us. Apart from that, when there is no one in a specific space, AI will turn off the AC automatically and turn on once someone comes back. Therefore we only consume energy where we really need it and deploy the energy depending on how much we need. To sum up, climate change and global warming are challenging all humans. Therefore Paris Agreement was adopted by 196 parties at COP21 in Paris. Implementation of it requires economic and social transformation which makes zero-carbon society an unavoidable trend [34].

1.4.2.4 Smart cities around the world

Compared to Society 5.0, smart city is a relatively simple concept. A smart city is a city that uses technology and data analytics to improve the quality of life for its citizens and visitors. A smart city uses data and technology to manage resources such as energy, transportation, and waste more efficiently, and to create a more livable, safe, and sustainable environment. In Ref. [35], in order to determine the smartest cities around the world, Berrone et al. evaluated 181 cities in more than 80 countries. They introduce cities in motion index (CIMI) as indicator. The CIMI ranking is based on a number of factors, including economic performance, environmental quality, social cohesion, mobility and transportation, urban planning and design. According to the results of CIMI, representative smart cities are: London, San Francisco, Barcelona, Santander, Nice, and Padova. Let us take Barcelona as an example.

As depicted in Fig. 1−15, the architecture of smart city can be divided into sensing layer, transmission layer, data management layer, application

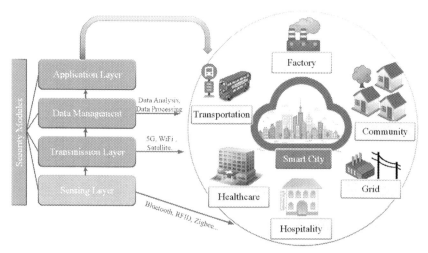

Figure 1−15 The architecture of the smart city.

layer, and security modules. Sensing layer is at the bottom of the structure, for it is used for harvesting all kinds of data from the environment with the aid of sensors and other devices. This layer consists of sensors and the technique for increasing sensing ability or efficiency in different contexts. For instance, in Barcelona, air and noise sensors are placed at intersections to track pollution levels. This information is shared publicly as open data. If the pollution readings at an intersection are high, the traffic signals are changed to allow vehicles to pass through without stopping, reducing the amount of fumes at the intersection.

After the data are gathered, how to send the data is determined by transmission layer which consists of short-range ICT technique like Bluetooth, Zigbee, near field communication (NFC), M2M, and RFID and transmission network technologies like 3G, 4G (LTE), 5G, and low-power wide area networks (LP-WAN). Data management is crucial for smart city operation because it affects the usage efficiency of data. The data management layer can be broken down into several subcategories: data fusion, data analysis, data processing, data storage, and event and decision management [36]. Finally, the processed data will be sent to application data to facilitate corresponding application. This layer interacts with citizens directly, hence vital to their user experience. In Barcelona, typical applications are smart parking system, smart lighting, smart waste management, and smart cycling [37]. Take smart parking system as an example, there is a lack of parking facilities there, so many drivers spend a lot of time looking for a space on the side of the road. To solve this problem, the city implemented a smart parking system. Sensors are installed in the asphalt and can detect whether a parking spot is occupied or not. Drivers can access this information through an app and see where the available spaces are located. These sensors have a battery and transmitter that sends signals to indicate the status of the parking spot. This information is displayed on a map in the app, so drivers can see it in real time.

1.4.3 Realizing Society 5.0

After introducing the basic concepts and theory about Society 5.0, the remaining question is how to realize it. The key solutions are as follows:

1.4.3.1 Habitat innovation

Habitat innovation refers to the development and implementation of new ideas, technologies, or approaches that enhance or improve the built environment in which people live and work. This can include a wide range of activities, such as designing and building sustainable and energy-efficient homes, creating walkable and livable communities, and developing technologies and infrastructure that support the needs of a growing population. Habitat innovation can be driven by a variety of factors, including population growth, changing demographic trends, advances in technology, and concerns about the environment and sustainability. The goal of habitat innovation is to create places that are more livable, efficient, and sustainable for people to live and work in.

To comprehend how habitat innovation works, we must come to realize that there are many different key performance indicators (KPIs) that can be used to measure the performance of a society. Some examples of KPIs that are commonly used to measure the performance of a society include the following:

- Economic indicators: These include measures such as GDP per capita, unemployment rate, and inflation rate, which can provide insight into a society's economic health and well-being.
- Social indicators: These include measures such as poverty rate, education levels, and healthcare access, which can provide insight into a society's social and cultural environment.
- Environmental indicators: These include measures such as air and water quality, resource depletion, and greenhouse gas emissions, which can provide insight into a society's environmental sustainability.
- Political indicators: These include measures such as voter turnout, political stability, and government effectiveness, which can provide insight into a society's political environment.
- Quality of life indicators: These include measures such as life expectancy, crime rates, and access to cultural and recreational opportunities, which can provide insight into a society's overall well-being.

We cannot single out one of these indicators as the only goal. For instance, if we would like to just increase GDP, we may end up becoming a society with mass production and mass consumption which compromises environmental indicators. All of these indicators must not be ignored. Otherwise, conflicts and crises will cause serious social unrest. A typical example is given by Ref. [31]. As shown in Fig. 1−16, to lower carbon emissions per capita, we can use structural transformation, technological innovation to maintain people's quality of life (QoL).

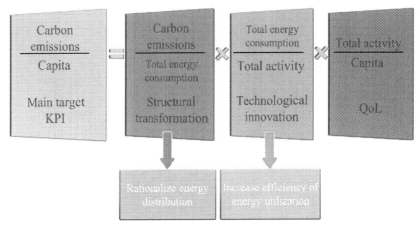

Figure 1—16 Formulae for calculating carbon emission [31].

1.4.3.2 Promoting regional revitalization

Regional uneven resources have led to a large number of people flocking to the center of big cities. They often have to bear high living costs and long commuting times but have no other alternatives. The expectation of regional revitalization is to increase infrastructure and industrial construction in urban noncentral areas or economically underdeveloped areas to attract more people to live in these areas, which not only improves their living conditions, but also reduces traffic jams and problems such as high maintenance costs caused by low infrastructure utilization in other regions.

It is true that the gathering of many enterprises in big cities has provided more job opportunities and forced young people to work here. This phenomenon has certainly improved the efficiency of enterprise operation and the ability of enterprises to cooperate and share resources, but when unbalanced development of regions is already very obvious, regional revitalization and government structural reform are necessary. Only when all regions develop in a coordinated manner and form their own distinctive economic and development models can the government establish strong political leadership and implement institutional changes to support a major shift in Society 5.0 strategy.

1.4.3.3 Industry—academia—government collaboration

Industry—academia—government collaboration, also known as "triple helix" collaboration, refers to the cooperation between these three sectors in order to advance research and development, stimulate innovation, and solve complex problems. In the context of Society 5.0, this type of collaboration can be particularly important in helping to realize the vision of a more advanced and sustainable society. Some specific ways that industry—academia—government collaboration can contribute to Society 5.0 include the following:

- Developing new technologies: By working together, industry, academia, and government can accelerate the development and deployment of new technologies, such as AI, the IoT, and renewable energy, that are necessary for Society 5.0.
- Solving complex problems: By pooling their expertise and resources, these sectors can work together to address complex global challenges such as climate change, aging populations, and resource depletion, which will be crucial for achieving the goals of Society 5.0.
- Promoting innovation: Collaboration between industry, academia, and government can foster a culture of innovation and encourage the development of new ideas and solutions that can drive progress toward Society 5.0.
- Facilitating the transfer of knowledge and technology: Collaboration can help to facilitate the transfer of knowledge and technology between the three sectors, enabling new technologies and ideas to be more quickly adopted and put into practice.

References

[1] L. Cheng, T. Yu, A new generation of AI: a review and perspective on machine learning technologies applied to smart energy and electric power systems, Int. J. Energy Res. 43 (2019) 1928−1973.
[2] T. Ahmad, D. Zhang, C. Huang, H. Zhang, N. Dai, Y. Song, et al., Artificial intelligence in sustainable energy industry: status quo, challenges and opportunities, J. Clean. Prod. 289 (2021) 125834.
[3] T. Ahmad, H. Zhu, D. Zhang, R. Tariq, A. Bassam, F. Ullah, et al., Energetics systems and artificial intelligence: applications of industry 4.0, Energy Rep. 8 (2022) 334−361.
[4] S.K. Jha, J. Bilalovic, A. Jha, N. Patel, H. Zhang, Renewable energy: present research and future scope of artificial intelligence, Renew. Sustain. Energy Rev. 77 (2017) 297−317.
[5] M. Luisa, D. Silvestre, S. Favuzza, E.R. Sanseverino, G. Zizzo, How decarbonization, digitalization and decentralization are changing key power infrastructures, Renew. Sustain. Energy Rev. 93 (2018) 483−498.
[6] Harnessing artificial intelligence to accelerate the energy transition, World Economic Forum, 2021.
[7] Digitalization and Energy, International Energy Agency, 2017.
[8] Cyber security threats in energy sector: everything you need to know. Available from: https://swisscyberinstitute.com/blog/all-you-need-to-know-about-cyber-security-threats-in-energy-sector/.

[9] S. Goel, Y. Hong, V. Papakonstantinou, D. Kloza, Smart Grid Security, Springer, 2015.

[10] R.E.H. Sims, H.-H. Rogner, K. Gregory, Carbon emission and mitigation cost comparisons between fossil fuel, nuclear and renewable energy resources for electricity generation, Energy Policy 31 (13) (2003) 1315−1326.

[11] F. Wang, et al., Technologies and perspectives for achieving carbon neutrality, Innovation 2 (4) (2001) 100180.

[12] J.M. Chen, Carbon neutrality: toward a sustainable future, Innovation 2 (3) (2021) 100127.

[13] G. Berndes, et al., Forest biomass, carbon neutrality and climate change mitigation, Sci. Policy 3 (7) (2016).

[14] P. Agreement, "Report of the conference of the parties to the united nations framework convention on climate change," (2022). Available from: https://unfccc.int/files/essential_background/convention/application/pdf/english_paris_agreement.pdf.

[15] H. Cheng, Future earth and sustainable developments, Innovation 1 (3) (2020) 100055.

[16] C. Hepburn, Y. Qi, N. Stern, B. Ward, C. Xie, D. Zenghelis, Towards carbon neutrality and China's 14th Five-Year Plan: clean energy transition, sustainable urban development, and investment priorities, Environ. Sci. Ecotechnol. 8 (2021) 100130.

[17] J. Pedersen et al., The Road Towards Carbon Neutrality in the Different Nordic Countries, (2022). Available from: http://doi.org/10.6027/temanord2020-527.

[18] M.-T. Huang, P.-M. Zhai, Achieving Paris Agreement temperature goals requires carbon neutrality by middle century with far-reaching transitions in the whole society, Adv. Clim. Change Res. 12 (2) (2021) 281−286.

[19] A. Osmani, J. Zhang, V. Gonela, I. Awudu, Electricity generation from renewables in the United States: resource potential, current usage, technical status, challenges, strategies, policies, and future directions, Renew. Sustain. Energy Rev. 24 (2013) 454−472.

[20] M. Mitchell, M. Newman, Complex systems theory and evolution, Encycl. Evol. 1 (2002) 1−5.

[21] G. Strbac, Demand side management: benefits and challenges,", Energy Policy 36 (12) (2008) 4419−4426.

[22] C.E. Shannon, A mathematical theory of communication,", Bell Syst. Tech. J. 27 (3) (1948) 379−423.

[23] C.C. Chan, L. Jian, Correlation between energy and information,", J. Asian Electr. Veh. 11 (1) (2013) 1625−1634.

[24] C.C. Chan, F.C. Chan, D. Tu, Energy and information correlation: towards sustainable energy,", J. Int. Counc. Electr. Eng. 5 (1) (2015) 29−33.

[25] K. Kumar, D. Zindani, J.P. Davim, Industry 4.0-Developments Towards the Fourth Industrial Revolution, Springer, 2019.

[26] 4 Networks and 4 Flows Integration promotes Hong Kong's Smart City, The Hong Kong Institution of Engineering. Available from: http://www.hkengineer.org.hk/issue/vol49-aug2021/.

[27] 2025 Outlook: Industry 4.0/Digital Industry/Society 5.0, IEEE Future Directions. Available from: https://cmte.ieee.org/futuredirections/2021/11/03/2025-outlook-industry-4-0-digital-industry-society-5-0/.

[28] Tag Archives: Industry 5.0, IEEE Future Directions. Available from: https://cmte.ieee.org/futuredirections/tag/industry-5-0>.

[29] Enabling Technologies for Industry 5.0, European Commission. Available from: http://www.4bt.us/wp-content/uploads/2021/04/INDUSTRY-5.0.pdf.

[30] What is Society 5.0? Cabinet Office (2016). Available from: https://www8.cao.go.jp/cstp/english/society5_0/index.html.

[31] Y. Shiroishi, K. Uchiyama, N. Suzuki, Society 5.0: for human security and well-being, Computer 51 (7) (2018) 91−95.

[32] C. Bai, P. Dallasega, G. Orzes, et al., Industry 4.0 technologies assessment: a sustainability perspective, Int. J. Prod. Econ. 229 (2020) 107776.
[33] Y. Wang, C. Guo, X. Chen, et al., Carbon peak and carbon neutrality in China: goals, implementation path and prospects, China Geol. 4 (4) (2021) 720−746.
[34] The Paris Agreement, United Nations. Available from: https://www.un.org/en/climatechange/paris-agreement.
[35] IESE cities in motion index 2018. IESE Business School. Available from: https://blog.iese.edu/cities-challenges-and-management/2018/05/23/iese-cities-in-motion-index-2018/.
[36] B.N. Silva, M. Khan, K. Han, Towards sustainable smart cities: a review of trends, architectures, components, and open challenges in smart cities, Sustain. Cities Soc. 38 (2018) 697−713.
[37] Smart City Series: the Barcelona Experience, Zigurat Global Institute of Technology. Available from: https://www.e-zigurat.com/blog/en/smart-city-barcelona-experience/.

CHAPTER TWO

Human-centered Renaissance from energy digitization

Chunhua Liu[1], C.C. Chan[2], Kuo Feng[1] and George You Zhou[3]
[1]School of Energy and Environment, City University of Hong Kong, Hong Kong
[2]The University of Hong Kong, Hong Kong SAR, P.R. China
[3]National Institute of Clean and Low-Carbon Energy, Beijing, P.R. China

Abstract

Before the third industrial revolution, humans mainly controlled the operation of energy systems. Currently with the development of energy digitization, big data and artificial intelligence algorithms are widely applied to control the systems. People are liberated from labor, but the purpose of people-orientation should not be ignored. In the fourth industrial revolution in the future, the humanities network will be revived as the superstructure, forming a new "Four Networks and Four Flows." This chapter mainly introduces the phenomena, challenge, elements, and foundation of human-centered Renaissance from energy digitization.

Keywords: Energy digitization; human-centered Renaissance; industrial humanity; human network; philosophy in science and engineering

2.1 Renaissance phenomena: cyber-physical-social

2.1.1 Concept of industrial humanity

In the Four Networks, the humanistic network is an important part of the superstructure. Integration of Four Networks and Four Flows within the "human–cyber–physical" system facilitates the establishment of a novel production relationship. The general definition of humanity refers to the advancement and core parts of human culture, that is, advanced values and customs. Culture is the symbols, values, and customs shared by human beings or a nation or a group of people. Symbol is the foundation of culture, while value is the core of culture. Custom, including living, moral, and legal custom, is the main content of culture. The rapid development of mobile Internet and digital network has fundamentally

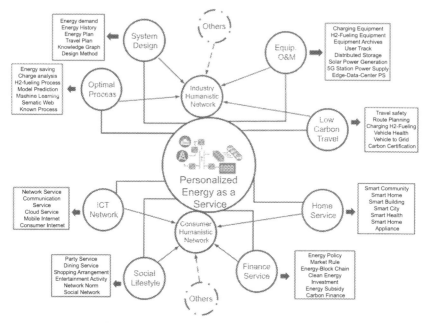

Figure 2—1 Personalized "Energy as a Service" driven by digital humanistic network.

changed people's lifestyles of shopping and socializing, forming a new type of digital consumer humanistic network. Fig. 2—1 is an illustration of personalized "Energy as a Service" driven by digital humanistic network.

In this chapter, a new concept is proposed and defined as industrial humanity. Industrial humanities are the advanced and core part of the industrial ecology, which includes the advanced business models and industrial standards in the industry. Industrial humanities, as shared by industrial participants, include industrial technology methods, industrial business models, and industrial standards. Industrial technology methods are the foundation of industrial humanity. Industrial business models are the core of industrial humanity. Industrial standards, including technical standards, safety requirements, legal rules, and domain knowledge, are the main content of industrial humanity. Driven by the Internet-of-Everything, the industrial humanistic network and the consumer humanistic network form a new digital humanistic network, which reflects the new production relationship between people, people-and-things, and things-and-things under the digital economy. How to understand the interaction and fusion significance of this new type of production relationship for new digital productivity is the key of the discussion in this book.

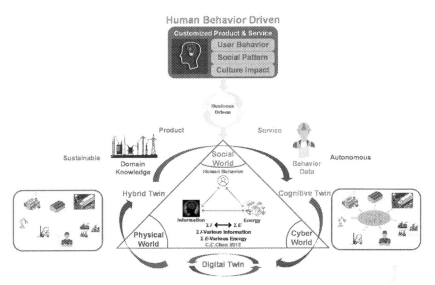

Figure 2—2 New production relationship formed by the human-cyber-physical system.

The essence of digital humanities is to study the new production relationship, as shown in Fig. 2—2, formed between human-cyber-physical systems under the digital economy. Taking the energy industry ecology as an example, it is the interaction between human behavior, information system and the physical system composed of energy/transportation, and the rules for establishing a smart energy system through this relationship. Digital humanities not only rely on the development of big data based on user behavior but also a need to build a more human-oriented cognitive intelligence to form the digital humanistic network. Because business models, energy policies, industry regulations, market rules, etc., are as important as user behavior, they reflect the integration of knowledge and data elements. This integration is a mandatory condition for the intelligentization of the energy system and is the core idea of this chapter.

2.1.2 Development of humanities network

For the energy revolution, a correct understanding of the new "human-cyber-physical" production relationship represented by the digital humanities network and its interaction for advocating the new digital productivity is crucial. With the in-depth development of the energy revolution, the fusion trend among digital humanity, information, energy, and transportation networks has gradually emerged, as well as developing as the core content of transportation (mobility) and information revolutions. The goal of fusion comes from the concept of combining two into one, that is, to obtain a greater value gain than simple superposition (the effect of $1 + 1 > 2$); its mechanism comes from the coupling of the system and the sharing of equipment, that is, to improve the energy utilization in the network as well as utilization of resources; its subversive nature is achieved by optimizing the dynamic causal relationship and complex connection logic between the fusion subjects through the digital humanity network; its economy benefit is driven continuously by the dynamic evolution of technology, industry to ecology between the subjects through the value flow.

The **4N4F** represents the fundamental change and influence of the fourth industrial revolution on the philosophical, scientific, and engineering levels of integrated thinking. It contains three major discoveries for the industry that integrates philosophical thinking, scientific theory, and engineering practice.

1. From philosophical level, it proves the value to establish a transition from simple linear thinking to global circular thinking, forming a value-added effect that is greater than the sum of the individual parts. The values of energy and materials flow can both be improved through coupling. This is the basic attribute of the sustainable development of the ecosystem and embodies the philosophy that the whole is greater than the partial superposition. That is to say, the partial transformation of a single energy flow or the partial use of a single material flow will reduce its own value, but if the two are coupled as a whole, the overall value can be improved.

2. From a scientific perspective, the **4N4F** reveals the internal mechanism of the interaction between energy, information, and human behavior and reflects the basic relationship between energy and information. Similar to Einstein's revealing the identity of mass and energy through the mass-energy equation, Academician Prof. C.C. Chan proposed and developed the idea of establishing a smart energy development path by studying the internal correlation between energy and information [1] in 2013. He also pointed out that human interaction (humanity network) is the key for energy/information carriers to become an intelligent fusion, and its typical scenario is the application of intelligent transportation of electrification and hydrogen energy, that is, the application of Internet-of-Vehicles and autonomous driving.

3. From an engineering perspective, it combines energy technology and information technology, establishes a smart energy operating system platform, and adopts an edge cloud architecture to combine the energy system and artificial intelligence/big data systems. Realize the integration of energy and information gridization and obtain the greatest benefits of industrial synergy. Its core idea is to operate the industrial joint innovation process of the Four Networks Integration, that is, to complete its own innovative research and development in a single technical field at the driver layer, focusing on data intelligence, and then forming a collaborative innovation integration of multiple technologies through the kernel layer, focusing on the knowledge intelligence, and then to go to the application layer to complete the ecological application innovation of related scenarios, focusing on cognitive intelligence. For example, the smart energy operating system is equivalent to the endogenous basic framework and rules of the sustainable development of the ecosystem and is a platform for the integration of human and machine intelligence. Due to the self-similar structure and characteristics of the smart energy operating system, the basic realization of human—machine intelligence integration can occur on the edge or the cloud side. As well as this, the smart energy operating system is a digital mapping and carrier of the integration of "human-cyber-physical" and an engine to realize smart energy.

2.1.3 Integration key of cyber-physical-social

The connotation of digitization in the current era is a data-centric ideological, theoretical, methodological, and technical architecture system, and its essence is to demand productivity from data. As far as digital transformation is concerned, digitalization is the direction and trend, and transformation is the path and means, and both are based on software- and service-driven changes, which inevitably accompany the reshaping of production relations. The key to the integration of the Four Networks is to establish the interactive integration of new production relations and productivity in the industrial revolution.

1. The integration of human-cyber-physical elements in the ecosystem is manifested in the integration of human-information-energy-transportation elements in the energy system. The intelligence of energy/transportation represents the new type of productivity formed under the digital economy, and the integration of human and cyber-

physical systems represents the new type of production relations required under the digital economy. Only by establishing a new type of production relationship between human and machine intelligence can it be possible to fully explore and release the huge productivity potential embedded in energy/transportation integration.

2. Because the integration of humans and machines can establish a new type of production relationship driven by data and knowledge, the new type of productivity under the digital economy has the possibility of sustainable development. This is because artificial intelligence that relies purely on big data lacks explaining ability and is unsustainable due to increased energy consumption.

The key to the integration of the Four Flows is to establish the integration of energy flow and material flow, which is also an inevitable demand and foundation for the establishment of a sustainable industrial ecology under the digital economy.

1. The goal of the digital economy is to continuously create new industrial value and accumulate huge social wealth through intelligence.

2. The digital economy forms an industrial ecological joint innovation model through the interconnection of information flows between different industries and fields to maximize industrial value.

3. The discovery of value requires the establishment of a new interconnection model of energy flow and material flow inside and outside the industry to minimize energy consumption and increase the utilization rate of material resources.

In addition, the development of 5G technology has brought the integration of networks, industries, and ecology to the comprehensive development of the digital economy at three levels. The advancement of 5G communication and data center technology has increased the need for information infrastructure construction at the network level. The high-speed, large-capacity, and low-latency characteristics of 5G technology can realize true interconnection among people, devices, and sensors, which provides the possibility for a comprehensive intelligent upgrade of industrial applications. Extended reality, as a next-generation mobile computing platform integrating human-cyber-physical, includes augmented reality, virtual reality, and mixed reality and is accelerating the integration of the real world and the digital world, thereby it will completely change the way human beings work, live, study, and play.

In the industrial era of past 200 years, competitive advantage is the effect of product scale, based on the direct communication between people. In the

Internet era in the past two decades, competitive advantage is the network effect, based on the online human–computer interaction. In the future era of Internet-of-Things, competitive advantage is the ecological fusion effect, based on the immersive human-cyber-physical integration.

2.1.4 Integration applications of humanistic network

The core of the Four Networks Integration is the integration of the humanistic network under the "human-cyber-physical" ecosystem, which embodies the top-level design thinking for the development of digital economy under the background of new infrastructure, that is, the method of establishing the integrated development of human-cyber-physical systems. The 4N4F integration based on human-cyber-physical framework is shown in Fig. 2–3. This is reflected in the applications of smart industry, smart energy, smart transportation, and smart cities. Among these, smart industry is the integration of the trinity of people, industrial information systems, and industrial physical systems; smart energy is the integration of the trinity of people, energy, and information; smart transportation is the integration of people, vehicles, and roads; and smart cities are the integration of the trinity of people, urban information systems, and urban

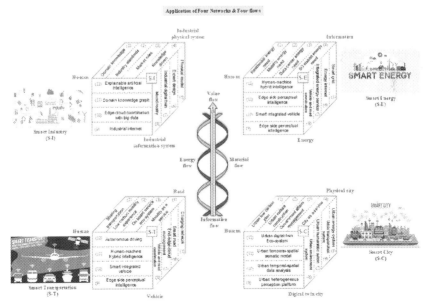

Figure 2–3 Four Networks and Four Flows integration based on human-cyber-physical framework.

physical systems. "Robot as a Service" is the most typical combination of artificial intelligence and Internet-of-Things technology and reflects the development trend of the Four Networks Integration of mobile functions, energy supply, edge intelligence, and human–cyber–physical interaction.

The large amount of industrial knowledge accumulated in the field for many years can solve qualitative problems well. However, in many scenarios, these mechanism models cannot accurately match the fluctuation of working conditions, and the industrial process is still a "black box." In addition, the large amount of tacit knowledge possessed by factory heads and craftsmen needs to be inherited and reproduced. To solve the above problems, industrial intelligence has become a new engine for intelligent empowerment of the industry. Through the intelligent cognitive engine driven by the knowledge graph, the intelligent prediction engine via the AI model, and the decision optimization engine through operations research and planning, the cyber–physical system-driven digital twin can evolve into a cognitive twin driven by the human–cyber–physical system, which allows many technologies, that were difficult to be applied in the past, to be embedded into industrial scenarios, and promotes the empowerment of industrial intelligence. Fig. 2–4 illustrates the smart industrial applications based on the evolution of human–cyber–physical systems.

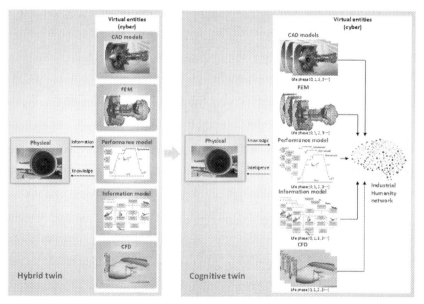

Figure 2–4 Smart industrial application based on the evolution of human-cyber-physical systems.

2.2 Renaissance challenge: human-artificial intelligence-environment

2.2.1 Background

In order to accelerate the development of artificial intelligence (AI) in the energy, information, and automobile transportation industries, establish the operating and interacting principles of the people-oriented human-information-physical network, and facilitate the realization of smart energy, industry, transportation, and city, the deep integration technology system with AI as the core plays a vital role. Great progress has been made in the development of AI technology in the past, the first-generation AI is based on knowledge-driven and the second on data-driven. Knowledge is usually produced by human experience in long-term production practice and social activities and applied to machine and intelligent systems. Furthermore, combined with AI, experience can be produced by data-driven AI autonomously. However, this kind of AI is unsustainable and unexplainable. For the fourth industrial revolution, AI will place more emphasis on its explaining ability and robustness. Hence, both the two kinds of AI cannot cope with the challenges to be faced in the fourth industrial revolution. To achieve this, we should take the best advantages of our humans, which means humans will play a key role in the development of the third-generation AI. Specifically, humans are the key to information integration and knowledge integration, thereby forming a new production relationship between people, AI, and the environment. All this awareness has far-reaching guiding significance.

To achieve the goal of "double carbon," we need to strive to transform our energy industry structure, among which the development of energy digital economy is an important strategic direction. At present, the development of China's digital economy has gradually reached a climax and has gradually become a new driving force for China's economic development. The adjustment of the energy industry structure with digitalization as the carrier and the promotion of low carbon, decentralized, and intelligent development of the energy industry is both an urgent matter and an irresistible trend.

The construction of digital infrastructure system based on AI technology is the cornerstone of the transformation of the energy industry. Digital infrastructure is an important prerequisite to ensure the safe storage of data resources in the process of digital energy transformation. Therefore we need to vigorously build an energy big data sharing platform based on AI technology, fully tap the cross-category and cross-business data collaboration management mechanism in the digital energy industry, break the "data monopoly" in the industry, and achieve data interconnection and intercommunication in the whole process of the energy industry. After building the above energy big data platform, we should also strengthen the rational governance of data resources, improve data utilization, and promote data resources to play a greater role in improving the efficiency of the energy industry. Among them, AI technology is a perfect bridge, which can connect massive data resources and various energy digital application scenarios and achieve personalized scheme customization under different energy digital application scenarios with low-carbon and efficient green indicators, including power generation, mining, logistics, travel, and other aspects. AI technology is more scientific and efficient in building the digital infrastructure of the energy industry and promoting the management and utilization of energy industry data. The construction of a data center is a typical application example. In the traditional mode, energy flow and information flow are independent, and there is much room for improvement. After multistream integration, the information flow provides an optimization strategy from a global perspective for the energy flow and carries out a real-time "physical examination" of the equipment in the energy flow link to achieve more reliable operation guarantee. In the process of data center operation, energy conservation and emission reduction are the most important assessment indicators. Through information flow technologies such as AI, the operation information of the data center is collected, analyzed, and dynamically modeled, and the optimal cooling strategy and power supply strategy are deduced in real time. At the same time, the overall energy efficiency of the data center is optimized by combining the computing power requirements of the upper data flow. In the daily operation and maintenance of the data center, AI can also be used to conduct predictive maintenance on key equipment in combination with operation data, so as to finally achieve the "automatic driving" of the data center.

In addition, it is also important to establish and improve the laws and regulations on the security of digital elements in the energy industry, improve the industry norms for the digital transformation of the energy industry, and ensure that there are laws and regulations to follow in the process of digital construction. The system is long term and fundamental and is an important foundation and premise to ensure the safe storage and safe operation of data elements in the process of energy digital transformation. Therefore we should strengthen the system protection mechanism of core technologies, such as data element security rating, hierarchical protection, data encryption, algorithms, transportation, backup, and other digital construction links, which are also the basis of energy digitization and the requirements of digital transformation. We need to provide solid legal support for the efficient and healthy operation of energy digitization. Taking the development of energy storage technology in the last decade as an example, the large fluctuations in the industry growth rate are not only related to the development of technology itself but also are affected by energy policies, industry norms, and market norms. The continuity, consistency, causality, and sustainability of energy policies, industry norms, and market norms can be further strengthened by establishing relevant top-level design and theoretical guidance in industrial joint innovation. In essence, energy policy, industry norms, and market norms are all direct impacts at the legal level on the development of the industry, so they can be understood as the result of the interaction between legal policy and technology. If this interaction process can be digitalized, that is, the key logic of energy policy, industry norms and market norms as well as domain knowledge for technological innovation and industrial development can be digitalized. AI technology can be used to integrate knowledge elements with existing data elements to fill in the shortcomings and weaknesses of new energy technology at the current legal level, which is conducive to joint innovation through the establishment of industrial ecology and could break through the bottleneck of industry development and jointly promote the energy revolution under the digital economy.

It is a great initiative to apply AI technology to the digital energy industry. As an important branch of the computer field, the core purpose of AI technology is to use machines to simulate human thinking process, and then replace people to complete the corresponding work, so as to realize the automation of intelligent behavior. From the perspective of the founding background, although the country has long proposed the "Internet + " smart energy development goal in 2015, the research on the intelligence of energy systems has a long history. For example, in the field of coal, in view of the long-standing problems faced by China's coal mining industry, such as mining difficulties and high degree of danger, domestic scientists put forward the concept of accurate coal mining supported by intelligent perception, intelligent control, big data, cloud computing, and the Internet-of-Things and affirmed the importance of AI technology in the coal mining industry. In addition, domestic scientists have conducted in-depth mining and analysis of coal mine data sets based on machine learning technology, assessed the possibility of coal mine power disaster, and made contributions to reducing the accident rate of coal mining. In addition, foreign scientists used AI models and decision tree analysis algorithms to build a coupled prediction model for predicting coal and gas accidents, reducing the accident rate. In addition, AI has also formed a great application prospect in the field of petroleum industry. The link process of traditional oil development includes the whole link process of exploration, transmission, exploitation, sales, and operation of oil and gas resources. AI technology can provide intelligent services for the design and maintenance of stone mining equipment, intelligent diagnosis and safety early warning of oil transmission equipment, prediction and optimization of oil exploration schemes, optimization of the layout of oil transport channels, prediction and control of oil operation costs, and other related scenarios. For example, foreign scientists use the method of artificial neural network to diagnose and predict the erosion degree of the pipe string in oil mines and provide an intelligent maintenance program. Even though there is still no universal AI technology applicable to the whole process of oil and gas resource development, AI solutions for specific problems can still provide great help. Finally, AI is also widely used in power systems and smart grids. In terms of intelligent dispatching and optimal control of power system, based on the development status and existing problems of power system in China, domestic scientists put forward the concept of intelligent dispatching of power system based on multiagent and built the framework of intelligent dispatching system. Other scientists introduced the concept of migration learning into the power system dispatching system to realize the automatic dynamic distribution of power generation control power through the analysis of historical data. Therefore in the context of the smart grid, it is of great significance to study the technologies of power grid system optimal transmission, automatic control, and risk assessment based on AI technology. In addition, AI technology is changing with each passing day, and the continuous improvement of intelligent algorithms has also indirectly promoted the upgrading of the energy industry. Taking the digitalization of knowledge map technology in coal mine energy as an example, the following three basic problems can be solved by establishing a coal mine knowledge computing model: (1) The information association level is not clear, the method based on "rules" only establishes the "image" association state between data and does not carry out in-depth and effective mining. The problems such as difficult to predict mine production, difficult to monitor, low efficiency, and many safety accidents cannot be effectively solved. How to realize the coupling and correlation of large-scale, multilevel, and nonlinear time and space information in mines and support the safe and efficient mining of resources has become a difficult problem in the development process of engineering field. (2) At this stage, the architecture is still based on data acquisition of digital mine, rather than data utilization. Smart mines lack a systematic information association mechanism but only a discovery strategy and a unified logical model and representation method. The existing architecture of digital mines has more evolved into a basic support system for smart services. (3) Lack of intelligent decision-making basis and effective control methods. Using modern mechanical, electrical, and AI technology to solve the problem of mine equipment control and realize intelligent unmanned mining is one of the core goals of the development of intelligent coal mines. Although it has broken through a number of key technologies, such as hydraulic support, remote control, and one button start stop, it has realized the coordinated automatic operation of equipment under simple and low-quality conditions. However, these technologies are mainly technical breakthroughs made within the mining system or in a single link, which cannot be combined to achieve continuous and stable mining under more complex geological conditions. It is necessary to carry out fusion and integration research of multiple technologies from the system level to solve the problems of insufficient data utilization, irrelevant information, and unintelligent control, and fundamentally improve the mining level, production efficiency, and the safety assurance ability of personnel and equipment of the coal mine. Another AI application scenario is in smart power generation. Taking the thermal power operation optimization process of smart power generation as an example, using the data analysis method of deep learning can optimize the prediction and simulation of boiler combustion near the operation point to a certain extent, but for a wider operation range, it requires a longer data preprocessing process and a large amount of parameter adjustment time costs. The effect often achieved is difficult to meet the requirements of the whole range, and there are unexplained safety risks in the control behavior. Therefore based on the introduction of the latest AI technologies such as knowledge atlas and semantic web, static knowledge such as design manual and artificial experience can be combined with dynamic operation data to learn, acquire, and generate dynamic knowledge, and form knowledge computing platforms that can evolve gradually through iterative upgrading at power plant level and group level, respectively, providing a scientific guarantee for the intelligent decision-making of the thermal power group. Among them, the digital twin of traditional thermal power plants will be upgraded to the integration twin and will further form the evolution twin.

In addition to the above traditional energy fields, AI technology has great potential in promoting the development of new industrial technologies. For example, we can integrate the end substation, 5G base station, charging and changing station (hydrogen refueling station), energy storage station and data center in the industrial park, promote more miniaturized industrial Internet multistation integration applications, and realize the coconstruction, cogovernance, sharing, and win–win of the four-network integration mode in smart energy. This superfusion station of intelligent industrial park integrating AI technology can achieve coordination and the mutual encouragement of energy flow and information flow. On the one hand, power system + new energy + energy storage can provide reliable and safe energy guarantee for 5G base stations and data centers and support data services, such as connection, sensing, and computing. On the other hand, the 5G base station and data center provide the energy system with the perception, circulation, calculation, and judgment of energy supply, power demand, and objective environmental information and provide more advantageous scheduling logic and performance optimization for the energy system to achieve mutual benefit and win–win results of energy flow and information flow.

2.2.2 Challenges of artificial intelligence in the Renaissance

As mentioned before, data-driven AI technology has now been widely developed in various scenarios. From the perspective of energy networks, AI is everywhere in smart energy and electric power systems today. AI 2.0 liberates people from the work of massive data analysis, prediction, judgment, and decision-making [1]. This technology can replace some human beings who have considerable work with significantly dull data. It also brings a huge increase in productivity because the workforce is no longer being overwhelmed by boring tasks. In addition, due to the powerful learning ability of AI itself, it can integrate information from different time and space dimensions, so that AI can not only help people by the liberation of their hands and brains but can also make better adjustments and decisions than human beings. As a result, these abilities facilitate the energy network to run more efficiently and directly create economic benefits for people.

Although from the above example, AI has brought many benefits to people and society; in fact, the current development of AI has also encountered certain bottlenecks and challenges. The fourth industrial revolution has brought unprecedented development opportunities and challenges to China because the fourth industrial revolution is to solve how to combine the innovation of emerging technologies with practical application scenarios to reshape productivity and production relations. At the same time, it also poses the issue of how to deeply integrate the digital economy and the real economy. Therefore in this context, the essential problem of the energy revolution under the digital economy has been transformed into the integration of AI and energy technology.

In the context of the country, how to apply AI to accelerate economic development is a long-term topic. Often the development and utilization of a certain technology need to rely on national policies and large-scale infrastructure construction. In 2020 China clearly proposed to vigorously develop the integrated infrastructure construction of smart energy, which further promoted the cross-technology integration in the energy field. For human users, our needs are diversified and personalized, so such challenges correspondingly put forward three requirements for the development of the energy industry: the first is that smart energy integration infrastructure needs to be highly integrated with information facilities; the second is that the energy technology innovation process needs to be matched with the update of industrial policies; and the third is that the business model of the energy industry needs to constantly adapt to the market rules in the reform. Under such development requirements, AI plays an important role in the convergence of Four Networks and Four Flows. As can be seen from the figure below, AI, as an indispensable technology in multiple links, mainly reflects its role and influence in the two aspects of digital characteristics and cross-industry characteristics in the convergence mechanism of Four Networks and Four Flows (Fig. 2−5).

The scientific foundation relied on by the formation of the Four Networks and Four Flows is the digital productivity generated by the integration of data and knowledge, and it is also the key to the breakthrough of the fourth industrial revolution in the application of AI in the industrial field. In the previous applications of industrial scenario, the amount of data

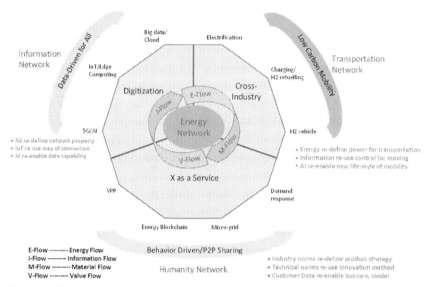

Figure 2–5 Basic characteristics of the integration of the Four Networks and Four Flows.

that can be generated is very limited, and the "cloud service + big data" model of traditional consumer Internet is also unsustainable, which directly leads to the unsustainability of the second-generation AI mentioned above. Therefore the model based on edge computing, small data, and big knowledge will become the mainstream model of convergence. And the "big knowledge" comes from the Industrial Humanities Network.

Since this model can reduce the dependence of machine learning samples on big data, the information network integrating knowledge and intelligent technology can become greener, more efficient and safer, and it can also more easily realize the integration of the decentralized distributed clean energy network and the new energy green transportation network. Furthermore, this will help to build the four-network convergence model that helps the fourth industrial revolution to release huge digital productivity. From an engineering perspective, the idea of combining energy technology and information technology, establishing a smart energy operating system platform, combining energy systems with AI and big data system, and attaching importance to data intelligence and cognitive intelligence is the core strategy of convergence of Four Networks and Four Flows.

On the other hand, attention should also be paid to the challenges for AI itself. In data-driven AI, one of the critical issues is the lack of complete explainability of big data algorithms. A good solution is arrived at and low-level parameters are generated but the embedded knowledge so created cannot be fully explained like autonomous intelligence. It is therefore difficult for most high-safety industrial applications to realize closed-loop decision-making processes, which in return greatly limits the in-depth application of AI in industry. Furthermore, there are a large number of cross-disciplinary industrial policies and technical specifications, and their update cycles often lag behind the pace of technological innovation. Due to the lack of timely traceable correlation for logic and causality between each other, industrial advancement is affected to a certain extent and has become a bottleneck encountered in the current industrial Internet and energy Internet applications. The third generation of AI is thus important, and it is essential to combine knowledge-driven and data-driven with knowledge explaining ability in the future development and applications of AI. This knowledge intelligence, as formed from the expert knowledge/industry standard, is to realize the integration with machine intelligence based on big data. Its significance lies in the fact that the technical framework and methods of human−machine hybrid intelligence can be established, which provides the possibility to achieve technological breakthroughs in explainable AI.

Faced with the above challenges, we must have a certain understanding of the key issues of the convergence of four networks and four streams. In fact, it is not difficult to find from the above elaboration that the key to how to reflect the role of humans and AI in the convergence of Four Networks and Four Flows is to solve the problems of sustainable development encountered in the process of breaking through the intelligentization of energy and transportation. One is the explainability and autonomy of AI, and the other is the sustainable energy demand of AI. Specifically, it can be divided into three points. First, the convergence of Four Networks and Four Flows can solve the explainability problem of AI through the integration of perceptual intelligence and cognitive intelligence and realize autonomous learning by continuously updating data intelligence into knowledge intelligence. Second, the big data requirements of the convergence of Four Networks and Four Flows can be reduced through the integration of data intelligence and knowledge intelligence. Besides, based on the framework of edge computing, small data, and big knowledge, the energy consumption demand of the processing of big data can be decreased further by integrating decentralized clean-energy/low-carbon transportation, finally, achieving the sustainable application of AI in the future. As a last thing, the future production relationship also needs to be reshaped, so it is necessary to have corresponding methodological support to reduce complex problems through the method of dimensionality reduction and decoupling and then ensure the effective realization of the transformation.

2.2.3 Industrial humanity in the Renaissance

The general definition of humanity refers to the advancement and core parts of human culture, that is, advanced values and custom. Culture is the symbols, values, and customs shared by human beings or a nation or a group of people. Symbol is the foundation of culture, while value is the core of culture. Custom, including living custom, moral custom and legal custom, is the main content of culture. The rapid development of mobile Internet and digital social network has fundamentally changed people's lifestyles of shopping and socializing, forming a new type of digital consumer humanistic network. Industrial humanity is a new concept. Industrial humanities are the advanced and core part of the industrial ecology, which includes the advanced business models and industrial standards in the industry. Industrial humanities, as shared by industrial participants, include industrial technology methods, business models, and standards. Industrial technology methods are the foundation of industrial humanity. Industrial business models are the core of industrial humanity. Industrial standards, including technical standards, safety requirements, legal rules, and domain knowledge, are the main content of industrial humanity.

The essence of digital humanities is to study the new production relationship formed between human-cyber-physical systems under the digital economy. Digital humanities not only rely on the development of big data based on user behavior but also need to build a more human-oriented cognitive intelligence to form the digital humanistic network. Because business models, energy policies, industry regulations, market rules, etc. are as important as user behavior, they reflect the integration of knowledge and data elements. This integration is a mandatory condition for the intelligentization of the energy system.

At present, the energy industry is making the transition from big data intelligence to cognitive intelligence. It needs to overcome the barrier from traditional perception intelligence to future decision intelligence. Due to the lack of complete explainability of big data algorithms, it is difficult for most high-safety industrial applications to realize closed-loop

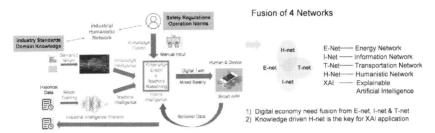

Figure 2−6 Convergence of four networks with industrial humanities network.

decision-making processes, which in return greatly limits the in-depth application of AI in the industry. In addition, there are a large number of cross-disciplinary industrial policies and technical specifications in the energy industry, and their update cycles often lags behind the pace of technological innovation. Due to the lack of timely traceable correlation for logic and causality between each other, it also affects the industry advancement to a certain extent, which has become a bottleneck encountered in the current industrial Internet and energy Internet applications.

The convergence of Four Networks with Industrial Humanities Network is shown in Fig. 2−6. Through the convergence of Four Networks and the previously proposed establishment concept of related industrial humanistic networks, the technical specifications, domain knowledge, safety specifications, and management standards, through semantic models, can be automatically transformed into a knowledge spectrum and a knowledge base. This knowledge intelligence, as formed from expert knowledge/industry standard, is to realize the integration with machine intelligence based on big data. Its significance lies in the fact that the technical framework and methods of human−machine hybrid intelligence can be established, which provides the possibility to achieve technological breakthroughs in explainable AI. Such framework and methods can point out the direction for the establishment of autonomous intelligence through the conversion of data to knowledge.

2.3 Renaissance elements: human-machine-things

Humanity is the advanced and core part of human culture, that is, progressive norms. Culture is the symbols, values, and norms common to human beings, people, or a group of people. Symbols are the basis of culture, values are the core of culture, and norms, including customary norms, moral norms, and legal norms, are the main content of culture. The industrial humanities are the advanced and core part of the industrial ecology, that is, the progressive business model and its industrial norms in the industry. Industrial culture is the industrial technology methods, industrial business models, and industrial norms that are common among industrial participants. Industrial technology methods are the basis of industrial culture, the industrial business model is the core of industrial culture, and industrial norms, including technical, safety, and legal norms, together with domain knowledge, are the main contents of industrial culture. Industrial humanities are the industrial technology methods, business models, and norms shared by industrial participants. The industrial technology method is the foundation of industrial humanities, the industrial business model is the core of industrial humanities, and industrial norms, including technical, safety, and legal norms, together with domain knowledge, are the main contents of industrial humanities. Driven by the Internet of Everything, the industrial humanities network and the consumer humanities network form a new digital humanities network, reflecting the new production relationship between people and people, people and things, and things and things in the digital economy.

2.3.1 The definition of human-machine-thing

The humanistic network is built on traditional social interaction, that is, offline cooperation and communication between people. Information is conveyed using trust. The information-based human network in the Internet era transfers offline collaboration and communication to online. The human—computer interaction method has changed the mode of human-to-human cooperation and trust. Human-machine-thing is an essential component of the human network in the current stage of the Internet of Everything. The so-called human-machine-thing combination relies on the industrialized human network and adopts the participant experience connection to realize the integration of human—machine intelligence. The industrialized human network of the Internet of Everything era combines the digital twin of the scene with the immersive experience of people to produce the disruptive model of human-machine-thing. Here, human refers to the human network, mainly referring to human behavior, policies, industry norms, or business models in

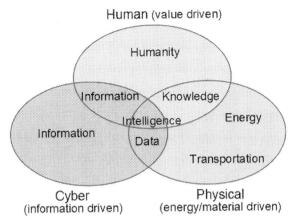

Figure 2-7 Four Networks and Four Flows in the digital economy.

the new energy environment; machine refers to the information network and information flow drive, mainly the flow of information and data in green energy applications; thing refers to the energy flow or material flow drive. Items contain energy and transportation networks, the material basis of natural society, and the solid foundation of human and information networks. In addition, the widespread use of artificial intelligence is now helping information networks to play a more critical role in the digital energy field. In the context of the 5G era, the massive amount of ultra-high speed, ed, and ultralow latency supported by 5G technology inject new vitality into mobile innovation. With the support of high-speed information transmission technology, human–machine–thing integration will enter a new stage (Fig. 2-7).

2.3.2 The importance of the human network in human-machine-thing

The energy revolution changes the shape of the physical world, the information revolution changes the order of the physical world, and the transportation revolution changes the connection of the physical world. On the other hand, humanity is the critical link connecting energy, information, and transportation. Under the guidance of people-oriented thinking, energy, communication, and transportation elements are fully applied to the construction of intelligent society, thus realizing a human-friendly social form in which humanities guide material development and materials return to humanities construction. From the perspective of innovation, integrating the industrial humanities network can establish direct interconnection between upstream and downstream of the industry from an industrial knowledge base and provide necessary ideas, methods, modes, and platforms for forming joint industrial innovation. From the business model perspective, it is possible to study user behavior and habits to meet personalized needs through integrating industrial humanities networks, thus bringing new types of services and user experiences. It has generated a market scale growth effect through the recent market elements. From the perspective of industrial development, the humanities network is the superstructure, and the energy, transportation, and information networks are the economic foundations. The integration of the four networks reflects the interaction and integration between new productivity and new production relations in the digital economy, which can better promote and release the massive potential of digital productivity.

For the energy revolution, it is crucial to correctly understand the new human–machine–thing with the in-depth development of the energy revolution; the trend of interactive integration of the human network, information network, energy network, and transportation network has gradually manifested itself and developed into the core of both the transportation and information revolutions. The goal of integration stems from the concept of merging two into one, that is, gaining more significant value gain than a simple superposition. Ultimately, the energy revolution will continue to develop in the direction of low carbon, intelligence, and terminal energy electrification under humanity's guidance.

2.3.3 The network integration

In modern society, where material civilization is fully developed, the need for spiritual enlightenment in human society has become increasingly important. The core of the humanistic network is to fully coordinate the development of the energy and transportation networks under the guidance of human-oriented thought and with the information flow of the information network. It enables the material to be further applied to the production and life of human beings. The green economy and environmental protection issues are widely concern worldwide. Energy needs to be used with maximum efficiency under the role of a human network, which requires the construction of modern human-oriented intelligent cities. In the future era of Internet-of-Things, competitive advantage is the ecological fusion effect, based on the immersive

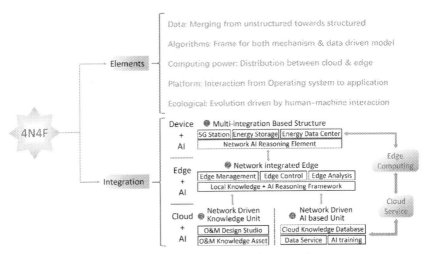

Figure 2—8 The basic elements of Four Networks and Four Flows integration.

human—cyber—physical integration. 4N4F integration includes five elements: data, algorithm, computing power, and ecological (Fig. 2—8).

1. Integration of data elements: Industrial big data has typical Internet-of-Things data characteristics, namely the rapid increase in the number of measurement points and the increasing frequency of data collection. The real-time database of traditional OT systems has gradually shown the lack of capability on horizontal expansion due to the outdated technical architecture, weak data analysis capabilities, and failure to support cloud service deployment, etc.; and traditional IT system relational databases will have various bottlenecks, such as low development efficiency, complex operation and maintenance, low operational efficiency, and slow application in launching to the market. Therefore it is necessary to adopt a technology fusion architecture with structured data management at the bottom level and unstructured data management at the top level to ensure the efficiency and economy of data management.

2. Integration of algorithm elements: the energy and transportation equipment layer in the Four Networks Integration industry has typical mechanism and physical law characteristics. In contrast, the application scheduling layer is closely related to user behavior and has specific data statistics features. Therefore in developing the Four Networks integration, industry needs to simultaneously pay attention to developing the mechanism and the data models. The final application algorithm must

be based on the data generated by the laws of physics, so it will reflect the profound fusion of algorithm application to ensure the explainability of the results.

3. Integration of computing power elements: Energy and transportation in the Four Networks Integration industry will generate massive amounts of real-time data. As time goes by, it will soon exceed traditional cloud computing platforms' growth rate and processing power. Edge computing has become a recognized development direction. The Four Networks converged infrastructure is based on the "Device-Edge-cloud" distributed architecture of the future 5G information network, so it depends to a large extent on the technological development and application level of edge computing. It must adopt a cloud–edge collaboration integration strategy in the distribution of computing power. Transferring cloud computing functions to the edge saves a large amount of data transmission and communication resources. This means that the critical needs of the industry's digitalization in agile processing, real-time business transaction, data optimization, application intelligence, security, and privacy protection can be met.

4. Integration of platform elements: The development of the Four Networks Integration industry reflects the deep integration of the traditional Internet and the new Internet-of-Things industry. The competitiveness model of enterprises has changed a lot. As value distribution among industry chains has surpassed traditional industries' boundaries, the conventional industry's asset-oriented advantage is gradually decreased by the competition pattern formed by the Internet platform strategy. This model has been verified many times in consumer and mobile Internet fields. The integration strategy of platform applications has gradually become the key to enterprises' competitiveness. From digital computers and automatic industrial control to mobile Internet intelligent phone applications and operating systems for various application scenarios launched by IT companies, they all reflect the importance of the platform strategy. Establishing AI ecological applications is the key to the platform application integration strategy. By selecting a universal development platform for AI developers, massive data preprocessing and semiautomatic labeling, large-scale distributed training, automated model generation, and device-edge-cloud model can deploy on-demand, helping users quickly create and adopt models and manage full-cycle AI workflows.

5. Integration of ecological elements: The purpose of the development of the Four Networks Integration industry is to establish an innovative industrial ecology to adapt to the future competitive business model driven by a new type of industrial socialization pattern. By establishing real-time, end-to-end, multidirectional communication and data sharing between people, products, systems, assets, and machines, each product and production process can be independently monitored to perceive and understand the surrounding environment. Via users and experts, brand, AI-defined content, continuous interaction, and self-learning with customers and the environment, valuable user experience can be created increasingly. Companies can also understand customers' individual needs in real-time and respond quickly. The changes that this data-based intelligence has brought to traditional industries are not only the improvement of production efficiency but also new product and service models derived from conventional products, opening brand-new business growth opportunities, while traditional industry operating models and competitiveness will be redefined. With the help of data generated by the Internet-of-Things, companies can provide customers with dynamic and personalized intelligent services. The essential difference between these services and traditional after-sales services is that the data collected through the Internet-of-Things can be used to analyze and predict customer needs more dynamically and systematically in real-time and continuously, and automatically optimize and adjust services based on the analysis results, and even can automatically adapt to the business environment, make independent decisions, and bring customers a highly personalized experience. At the same time, higher requirements are put forward for the credibility and security of data, and blockchain technology will solve the critical problem of data sharing. In addition, companies can further create new service models through the Internet-of-Things, such as opening their manufacturing capabilities to provide production services for other companies, providing C2B customized services according to customer needs, and providing customers with financing based on Internet-of-Things data with insurance services. Those industry leaders can also build platforms based on the Internet-of-Things and become the center of the industrial ecology.

2.3.4 Conclusion

In a nutshell, human-machine-thing is more of a multidisciplinary and multiobjective model of synergistic development. It reflects not only the decisive role of the economic base on the superstructure but also the influence of the human network as the superstructure on the energy and transportation networks. The development of science and technology ultimately acts on the progress of human society and improves people's living standards. Under the impetus of advanced humanistic concepts, the development of energy, transportation, and information will make great strides toward integrating each other, thus benefiting the whole human society. This is the unstoppable trend of the times. Just as the Renaissance opened the curtain of capitalism innovation for Europe, the guidance of the humanistic network will open the prosperity and development of modern and future cities. It will also bring more economic growth points for human society. This is not only a technological change but also an ideological change. Under the guidance of the Humanities Network, the development of all undertakings will continue to progress in the direction of integration, crossover, and sharing.

2.4 Renaissance foundation: philosophy-science-engineering

The formation of digital productivity through the integration of data and knowledge is the key to the fourth industrial revolution breaking through the application of artificial intelligence in the industrial field. Due to the limited amount of data in industrial scenarios, the widespread "cloud service + big data" model of the traditional consumer Internet, consuming too much energy and requiring massive data, is not sustainable for industry applications, and the model based on "Edge cloud + small data + big knowledge" will become the trend. The critical issues for the development of typical models of intelligent systems are not the upper and lower bounds that information considerations would provide. Rather it is the complex interdependence between the physical limitations of thermodynamic boundaries of energy transfer and the human dimensions of economic, social, and political decisions and regulations that is crucial to this discussion. The "big knowledge" comes exactly from the integration of the industrial humanistic network.

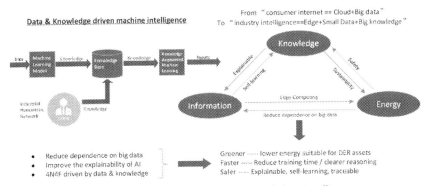

Figure 2-9 Integration of knowledge intelligence and data intelligence.

Fig. 2-9 shows the integration of knowledge intelligence and data intelligence. The dependence of machine learning samples on big data is reduced. With the introduction of the knowledge-based Industrial Humanity Network, the interpretability of AI is enhanced greatly. The information network integrating knowledge intelligence technology can be greener, more efficient, and safer, and, as a result, become more sustainable. The information networks are thus easier to integrate with decentralized distributed clean energy networks and new energy green transportation networks, forming a "Four Networks" fusion model, which will help the fourth industrial revolution unleash huge digital productivity.

The **4N4F** typically represents the fundamental change and influence of the fourth industrial revolution on the philosophical, scientific, and engineering levels of integrated thinking. It is a strategic industry innovation thinking that contains three major discoveries: interactively integrating philosophical thinking, scientific theory, and engineering practice.

2.4.1 Interactively integrating philosophical

From the philosophical level, the **4N4F** proves the value to transfer from simple linear thinking to global circular thinking, forming a value-added effect that is greater than the sum of the individual. Energy flow and material flow can convert to and improve both values through coupling. This is the basic attribute of the sustainable development of the ecosystem and embodies the philosophy that the whole is greater than the partial superposition. That is to say, the partial transformation of a single energy flow or the partial use of a single material flow will reduce its value, but if the two are coupled as a whole, the overall value can be improved.

- Through repeated iterations of the self-similar value conversion process of "energy–material–energy–material," sustainable circular ecology can be realized from the three dimensions of time/space/type, which is equivalent to using energy/material coupling means to solve the energy storage problem.
- Sustainable recycling ecology can continuously improve energy intelligence by transforming the local entropy increase process of the system into the overall exergy increase process, that is, the process of partial value declining can be turned into the overall value increase process, which can continuously improve energy intelligence, that is, the system converts useless resources into useful capabilities. In this method, the exergy is introduced to evaluate the intelligence of the whole system.
- Based on the self-similar network from nanogrid, microgrid, and distribution network to regional energy network, the 4N4F system method can simulate the self-organizing fractal structure of the natural ecosystem, that is, the mobile transportation network is similar to the animal community and the low-carbon building network mimics the plant community.
- H. Haken, the founder of synergetic, pointed out that the organization of a system can be divided into self-organization and hetero organization. Hetero organization refers to the organization formed by the external commands of a system. That is, the objective function acquires a specific value according to the external command. The specific method is to feedback on the output information and obtain the value of the objective function by adjusting the parameters. Self-organization means that the system forms an orderly structure automatically and harmoniously according to some rules of mutual understanding without external commands. In other words, self-organizing structures can evolve spontaneously and orderly, so it is an intelligent process. To get this ordered structure, the concept of entropy in thermodynamics is introduced to describe how ordered or chaotic a system is. Therefore self-organization can be regarded as a process in which a system continuously reduces its entropy content and improves its order degree by exchanging matter, energy, and information with the outside world. That is to say, more intelligent systems.

Through the exchange of energy and material, the continuous cycle of iteration and evolution progresses, and an urban smart energy ecosystem will finally be developed in this method. Fig. 2–10 shows the self-similar characteristics of the fractal energy network system.

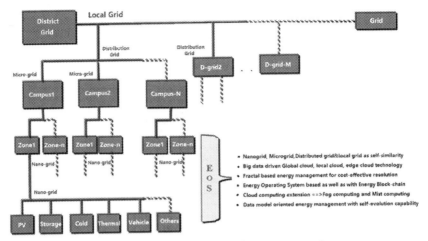

Figure 2−10 Self-similar characteristics of fractal energy network system.

2.4.2 Scientific theory

From a scientific perspective, the **4N4F** reveals the internal mechanism of the interaction between energy, information, and human behavior and reflects the basic relationship between energy and information. Fig. 2−11 shows the concept of Four Flows via energy storage entropy coupling energy and information. Similar to Einstein's revealing the identity of mass and energy through the mass-energy equation, Academician Prof. C.C. Chan proposed and developed the idea of establishing a smart energy development path [1] by studying the internal correlation between energy and information [2] in 2015. He also pointed out that human interaction (humanity network) is the key for energy/information carriers to become an intelligent fusion, and a typical scenario is the application of intelligent transportation of electrification and hydrogen energy, that is, the application of Internet-of-Vehicles and autonomous driving.

- To realize the value conversion of energy flow and material flow, there are two barriers to overcome. One is the entropy increase limitation of the second law of thermodynamics, and the other is how to ensure the sustainability of the system through continuous added value.
- Entropy is regarded as the key correlation concept between energy and information [1]. And the concept of storage entropy was introduced to show the entropy contribution caused by the interactive behavior between two systems and to quantify the energy intelligence level of an energy system. If using minimized information flow can

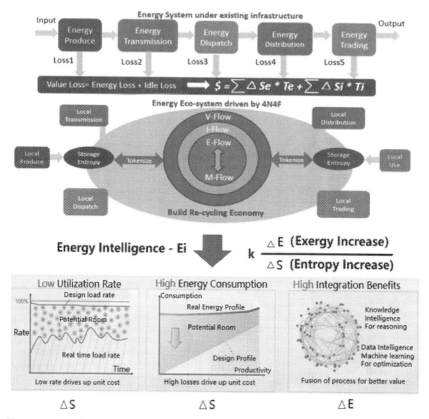

Figure 2–11 The concept of Four Flows via energy storage entropy coupling energy and information.

maximize storage entropy conversion into exergy energy, the system will have the highest energy intelligence and the highest capability to do work. Energy storage has been observed to show fractal structure at different temporal and spatial scales in a multienergy integrated system.

- The fundamental reality is that energy and matter are limited and thus must be allocated and distributed with consideration of values, justice, and equality, while information can be distributed and shared as widely as desired, leading to questions of access and education to understand and use it. It would not be simply a matter of producing more energy and deriving information but of changing the social use of energy and creating new business models and opportunities for innovation that are integrated more effectively with the understanding of the needs of the local and global society with limited resources [2].

- By coupling the information flow to establish the open characteristics of energy flow and material flow, the entropy increase of the closed local system can be transformed into the exergy increase of the open fusion system, that is, turning the useless energy into useful. Conventional local systems cannot operate optimally because unpredictable loads prevent the production sides from working at a high using rate and too much redundancy is required, which leads to entropy increase. By correlating different flows through the networks, an interactive system is constructed, and the exergy increase is realized through the optimal configuration of resources based on interpretable knowledge-based AI. The engineering implementation can be realized through value tokenization, using blockchain technologies to digitize the value of the information flow and facilitate the coupling of energy flow and material flow.
- The value gradient of each link of the industrial ecology is the key to driving the coupling of energy flow and material flow, and the value transformation between them always points to the direction of value gradient increasing.
- The value gradient of the industrial ecology is affected by industrial policies, industry norms, market rules, and field development, which will be included in the humanistic network and needs to be jointly promoted through industrial joint innovation. For example, the 14th Five-Year Plan and the long-range objectives through the year 2035 makes an extremely important role in the humanistic network and directing the value flow, influencing construction facilities related directly to the energy network and transportation network. The Paris Agreement and various acts of the United States also are significant, being key nodes in the digital humanistic network and determining the value flow. The report on the work of the government in China, and other results or feedback on the policies and regulations about the industry, again reinforce their influence on the flow of energy and material, and the network of energy and transportation. In the future stage, with machine learning techniques introduced, the value flow and information flow can be more intelligent, automatically evolving toward the value gradient and extracting value and information from the interactively integrated energy and material flow.

2.4.3 Engineering practice

From an engineering perspective, energy technology and information technology are combined, establishing a smart energy operating system platform and adopting an edge cloud architecture to combine the energy system and artificial intelligence/big data systems, therefore realizing the integration of energy gridization and information gridization and obtaining the greatest benefits from industrial synergy.

The core idea is to make the industrial joint innovation process of the Four Networks Integration become an operating system, that is, to complete its innovative research and development in a single technical field at the driver layer, focusing on data intelligence, and then forming a collaborative innovation integration of multiple technologies through the kernel layer, focusing on the knowledge intelligence, and then to the application layer to complete the ecological application innovation of related scenarios, focusing on cognitive intelligence.

Most distributed energy systems are not smart enough to be economically viable as we have not figured out whether we can or how to quantify an index level for energy intelligence. We can neither evaluate the behaviors between different microgrids nor between present and past status for the same microgrid. This has prevented the energy industry creating standard integration principles and evaluation rules at the system level.

Fig. 2–12 demonstrates the engineering application method based on the idea of the operating system.

- The advantage of the Four Networks Integration approach is that it can form a self-similar operating system structure at both the micro and macro levels, which is adaptive to the customization and fault tolerance requirements of the system and realizes the complex industrial ecological joint innovation under the optimal cost performance.

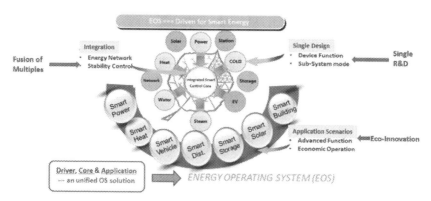

Figure 2–12 Engineering application method based on the idea of operating system.

- Information has been the key to the self-organized evolution of human society. Information technology has developed an Operating System-based software layer to manage hardware generically. An Energy Operating System framework could be developed for the successful integration of multienergy-based system operation.
- The result is that through the multilevel self-similar fusion of equipment, technology, network, industry, and ecology, the system's self-circulation evolution innovation process can be completed under the framework of decentralization, the application of artificial intelligence can be explained, and self-learning will be realized.
- The smart energy operating system is equivalent to the endogenous basic framework and rules of the sustainable development of the ecosystem. It is a platform integrating human and machine intelligence.
- Due to the self-similar structure and characteristics of the smart energy operating system, the basic realization of human—machine intelligence integration can occur on the edge of the cloud side. It has been proven to some extent [3,4] that, a self-organized intelligent network tends to grow into a fractal structure with self-similar and scale-free features.
- The smart energy operating system is a digital mapping and carrier of the integration of "human-cyber-physical," and an engine to realize smart energy. A software-defined-energy framework built upon the "operating system" through the four networks' integration can be implemented to support the future depicted.

References

[1] C.C. Chan, F.C. Chan, Tu Dan, Energy and information correlation: towards sustainable energy, J. Int. Counc. Electr. Eng. 5 (1) (2015) 29—33.
[2] Y. Zhou, C.C. Chan, D. Zhang, et al., Smart energy evolution road-map based on the correlation between energy and information, Energy Procedia 158 (2019) 3082—3087.
[3] C. Song, S. Havlin, H.A. Makse, Origins of fractality in the growth of complex networks, Nat. Phys. 2 (2006) 275—281.
[4] C. Song, S. Havlin, H.A. Makse, Self-similarity of complex networks, Nature 433 (2005) 392—395.

CHAPTER THREE

Integration of energy, transportation, and information with humanity

Xiaohua Li[1], C.C. Chan[2], Hang Zhao[3], Jin Li[1], Yin Yao[1] and George You Zhou[4]

[1]Electric Power Engineering, Shanghai University of Electric Power, Shanghai, P.R. China
[2]The University of Hong Kong, Hong Kong SAR, P.R. China
[3]Robotics and Autonomous Systems Thrust, The Hong Kong University of Science and Technology (Guangzhou), Guangzhou, P.R. China
[4]National Institute of Clean and Low-Carbon Energy, Beijing, P.R. China

Abstract

With the new stage of the sixth information revolution, the "four networks and four flows" concept is being applied based on the online human—computer interaction network. This chapter presents an overview of the progression of the information revolution and its amalgamation with the forthcoming Metaverse. Initially, the potential opportunities and challenges associated with the energy revolution are proposed. Subsequently, the transportation transformation during the era of humanistic digital transformation is deliberated upon. Finally, the pivotal role of the human network in intelligent applications is underscored.

Keywords: Four networks and four flows; information revolution; Metaverse; 5G technology; electrification; intelligent; smart energy; smart transportation; Industry 5.0; humanity network

3.1 Social behavior patterns under the information revolution

3.1.1 Development history and process of the information revolution

In the annals of human history, five distinct information revolutions have ushered in a great leap forward for human civilization. Presently, the sixth information revolution has progressed into a new stage.

Integration of Energy, Information, Transportation and Humanity.
DOI: https://doi.org/10.1016/B978-0-323-95521-8.00019-1
73

The advent of the information revolution has provided human beings with novel means of production, leading to the significant expansion of productive forces, organizational and management modes, and a transformation of industrial and economic structures [1]. The fundamental of the information revolution lies in the development of information technology, and the core connotation of the information revolution is intelligence. The information age has undergone several phases of digitalization, networking, and intelligence, and currently, it has entered the epoch of artificial intelligence.

3.1.1.1 The history of the information revolution

The first information revolution, which carried the weight of language, served as the bedrock for humanity to convert natural information into social information. The ensuing three information revolutions, characterized by the advent of papermaking and printing, the creation of telegraphy, telephony, and television, and the emergence of computers and the Internet, have also reshaped the structure of the social organization by transforming social thought processes. Following the industrial revolution, the thought process of human society has progressed toward scale, standardization, mechanization, and the division of labor as a consequence of the fourth information revolution, leading to the formation of various modern organizations.

The new information revolution includes the fifth and the sixth information revolution that happened at the end of the 20th century and at the beginning of the 21st century, separately. The fifth information revolution marks the point at which human society entered the information age from the industrial age, defined by the birth of the Internet [1]. The sixth information revolution promotes a transition from the information age to the intelligence age. This transition is taking place now, marked by the advent of big data, cloud computing, the Internet of things (IoT), intelligence, and other technologies. After the fifth information revolution, human society appeared to be in a subversive new situation. The Internet makes it possible to connect the world simultaneously, and society has entered a new era of comprehensive globalization. The emergence of computer technology directly promotes the advance of new occupations and changes in human economic and social forms; at the same time, Internet thinking is also subtly affecting the thinking mode of human society.

3.1.1.2 The development of the sixth information revolution

Since the beginning of the 21st century, a new generation of information technology has been emerging, including cloud computing, the IoT, big data, and mobile Internet, all of which are the core of a new stage of the sixth information revolution, which human society has entered.

Specifically, cloud computing refactors the structure of the whole IT industry from large to powerful. The IoT integrates intelligent perception, recognition technology, ubiquitous computing, and ubiquitous networks. It is called the third wave of the development of the information industry worldwide after the computer and the Internet. It will be the next "important production" to promote the rapid growth of the world and a new round of lifestyle changes. Big data can help open a new management mode filled with numbers to better concentrate on a "big society." It will profoundly impact the future of the economy, politics, culture, and other aspects. With the development of global intelligence, interconnection, and mobility, mobile Internet technology will be rapidly developed and widely used.

3.1.1.3 The next stage of the Internet—information Internet to the Metaverse

The Metaverse integrates virtual and physical universes, an important direction for future Internet upgrades. It combines digital and reality so that users can map various activities in the real world to the virtual space. The key technologies of the Metaverse include digital twins, blockchain, interaction, 5G network, cloud computing, artificial intelligence, and the IoT. Among them, the IoT provides an essential source of information; interaction offers an immersive experience; blockchain builds a virtual and real economic system; cloud computing and 5G networks are essential infrastructure constructions; artificial intelligence technology improves the wisdom of the Metaverse. These factors help realize the concept of transcending the universe [2].

The Metaverse is also inseparable from the smart city. The interaction between the physical world and the Metaverse is the key to the operation of the meta–smart city [3]. Only Web 3.0 is capable of such an essential task among the current technologies. Strictly speaking, Web 3.0 is not a single technology but a combination of several technologies. To date, Web 3.0 still needs to be clearly defined but has generally accepted characteristics:

1. The network of Web 3.0 involves a wide range of intelligent devices and users, the vast majority of whom can collect information from the

physical world and send data to the Metaverse to build a virtual city. At the same time, Web 3.0 can also pass the data of the Metaverse to the physical world.

2. It will be more intelligent depending on the smart brain in the meta-smart city and the data from the physical world. The two parallel worlds will influence each other in continuous interaction and coevolution.

3. Web 3.0 will become a trusted decentralized network. Based on blockchain technology, Web 3.0 can maintain the reliability and security of user identity and data. It will use the InterPlanetary File System to replace the current HTTP (Hypertext Transfer Protocol) protocol, thus achieving decentralization (Fig. 3−1).

3.1.2 Main technologies of the contemporary information revolution

3.1.2.1 Contemporary development and application of artificial intelligence technology

The formation of digital productivity through the integration of data and knowledge is the key to the fourth industrial revolution to break through the application of artificial intelligence in the industrial field. The conventional "cloud service + big data" framework utilized by the consumer Internet is unsustainable in industrial settings, primarily due to the limited amount of data available. In contrast, the "Edge cloud + small data + big

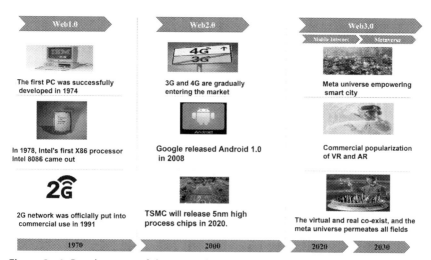

Figure 3−1 Development of the network.

knowledge" model is deemed as the trendsetter. The "big knowledge" comes from integrating the industrial humanistic network (Fig. 3—2).

Thanks to the ability to reduce the dependence on machine learning samples on big data, the information network integrating knowledge and data intelligence can be greener, more efficient, safer, and easier to integrate with decentralized distributed clean energy networks and new energy green transportation networks. The integration network will finally form a four networks for fusion (4N4F) model, which will help the fourth industrial revolution unleash tremendous digital productivity.

In 2020 Huawei teamed up with Sinopec and launched smart gas station business innovations to improve service quality, marketing efficiency, and safety monitoring capabilities in gas stations through AI and big data technology. Among those innovations, the "AI license plate payment" mobile phone app can recognize the license plate with AI and submit bills to customers in real-time. The APP also gives real-time information at the gas station about the customer queue, waiting time, payment status, and even updated efficiency information on fuel types, such as diesel and No. 98 gasoline. Supported by this data, the company can standardize the service experience of gas stations and promote the efficiency of accessing, passing, filling, and returning customer rates. Through data + AI, the company can also control tax evasion and other gray areas, creating a fair and transparent operation business environment. In the future, the progress of 4N4F integration will ensure a digital and intelligent promotion of industries like energy and transportation.

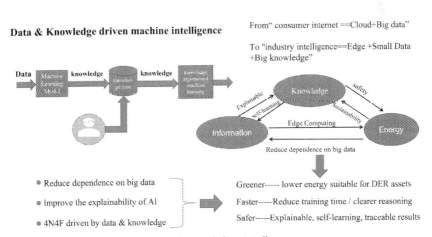

Figure 3—2 Integration of knowledge and data intelligence.

3.1.2.2 Contemporary development and application of 5G technology

The development of 5G technology has brought the integration of networks, industries, and ecology to help the comprehensive development of the digital economy at three levels.

First, the advancement of 5G communication and data center technology has increased the need for information infrastructure construction in forming an integrated network. Therefore Huawei has proposed an autonomous driving framework based on the network [3] and data center. According to various levels of human—machine integration, the framework defines the self-organizing intelligence path for future networks and data centers. Although derived from transportation, the autonomous driving mode significantly empowers the entire industry (Figs. 3—3 and 3—4).

The high-speed, large-capacity, and low-latency characteristics of 5G technology can realize actual interconnection among people, devices, and sensors. It also provides the possibility for a comprehensive intelligent upgrade of industrial applications. Extended reality (XR), as a next-generation mobile computing platform integrating human, cyber, and physical, includes augmented reality (AR), virtual reality (VR), and mixed reality (MR). XR is accelerating the integration of the real and digital worlds, completely changing how humans work, live, study, and take recreation.

The change brought by 5G is not simply the increased network speed and capacity. 5G will bring about a subversive industrial change. For example, the 5G network is essentially a self-driving network, which, in other words, is a new intelligent network system that can realize self-organization, optimization, learning, and upgrading. Fig. 3—5 is an illustration of the trend and benefits of 4N4F (Fig. 3—6).

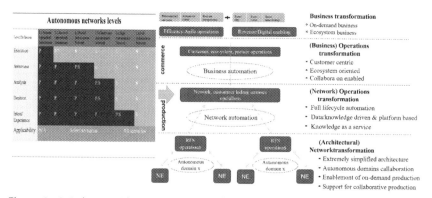

Figure 3—3 Industry-wide empowerment of autonomous driving mode.

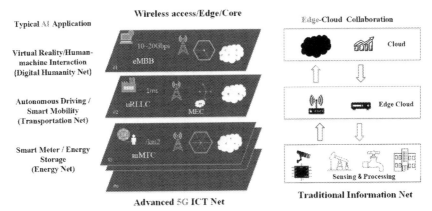

Figure 3−4 Industrial ecological innovation based on the integration of four networks via 5G/Cloud/AI.

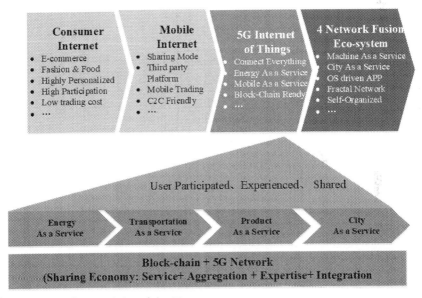

Figure 3−5 Characteristics of the Metaverse.

3.1.2.3 Development and application of future Metaverse technology

The application of information network technology is inevitable for developing the Metaverse because it is an entirely digital world. To build and realize the Metaverse, a higher degree of or even complete digitization must be achieved first [4] (Fig. 3−7).

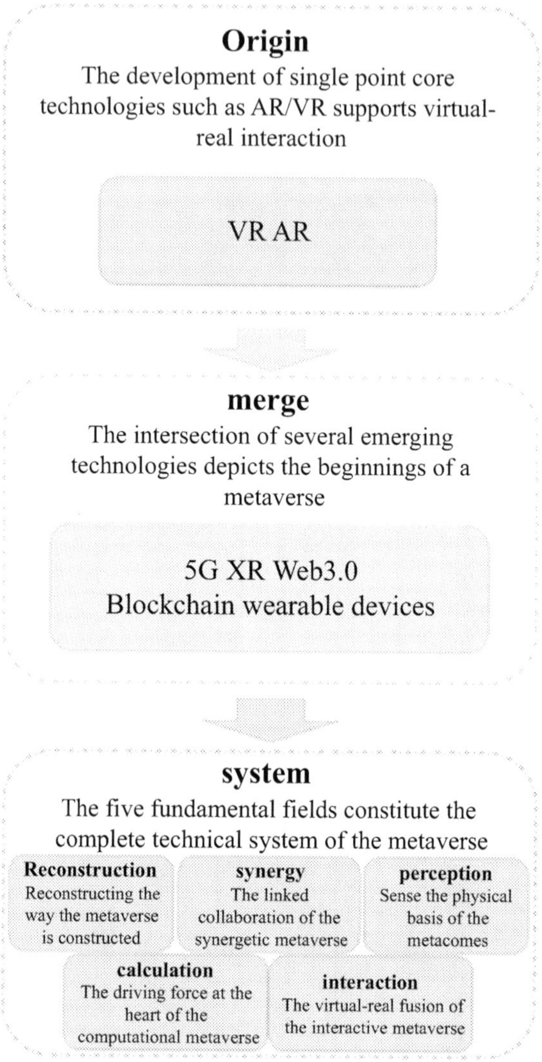

Origin

The development of single point core technologies such as AR/VR supports virtual-real interaction

VR AR

merge

The intersection of several emerging technologies depicts the beginnings of a metaverse

5G XR Web3.0
Blockchain wearable devices

system

The five fundamental fields constitute the complete technical system of the metaverse

Reconstruction	**synergy**	**perception**
Reconstructing the way the metaverse is constructed	The linked collaboration of the synergetic metaverse	Sense the physical basis of the metacomes

calculation	**interaction**
The driving force at the heart of the computational metaverse	The virtual-real fusion of the interactive metaverse

Figure 3–6 The development trend and benefits of 4N4F.

Application of artificial intelligence (AI) technology in the Metaverse: The AI cloud platform can provide solutions for two kinds of problems. The first is to provide the core atomic capabilities of AI for thousands of applications in the Metaverse to realize intensive reuse and sharing. The other is to provide scenario-oriented data structured processing, automatic annotation, and model training and reasoning [5] to govern the massive data in the Metaverse.

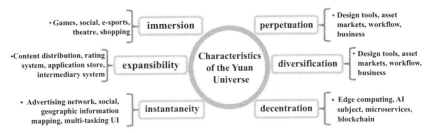

Figure 3—7 Technological evolution in the Metaverse.

Application of blockchain technology in the Metaverse: The Metaverse is closely related to blockchain because it naturally contains encrypted identities. The addresses associated with distributed ledgers can be directly used between address owners for mutual authentication, identity management, verification, and the underlying stack of transactions, all of which are the native capability of blockchain. Thus the only information available in the Metaverse is the same encrypted address for the blockchain identity of all entities [6].

Application of 5G technology in the Metaverse: 5G is not just a network access technology but should be seen as a set of technologies with 5G as the main body and in close contact with other modern technologies. Integrating 5G with XR, cloud computing, security, and other technologies will be one of the main driving forces for the progress of the Metaverse [7].

3.1.3 Integration and application of information network technology and the future Metaverse

Metaverse is a novel Internet application where the operation rules are stand-alone. At the same time, some elements are integrated with the real world by various technologies such as XR technology, blockchain, and digital twin [1,8].

The conceptual evolution of the Metaverse: In 1992 Neil Stephenson, a famous American science fiction writer, put forward the concept of the Metaverse in his novel Avalanche. This concept fulfills people's fantasies about a harmonious world in virtual form. The later Metaverse evolved from this concept to a future digital existence of human beings and a new digital lifestyle.

Technological evolution in the Metaverse: Single-point technologies such as AR or VR support the original virtual—real interaction. Then several emerging technologies fused to form the embryonic Metaverse. Finally, the five primary fields of the Metaverse are perception, computation, reconstruction, interaction, and collaboration. These fields work together to create a complete technical system for the Metaverse.

Characteristics of the Metaverse: The Metaverse has six features: immersion, expansion, real-time, sustainability, diversification, and decentralization, laying the foundation for the development potential and importance of the Metaverse in the future.

The strategic implications of the Metaverse: For human civilization, the Metaverse is based on the laws and principles of the physical world but diverges from the constraints and limitations of the physical world to upgrade and transform resources. By such means, the energy can reach maximum utilization, and the information can use highly efficient transmission so that the information technology will finish upgrading the physical world. For scientific research, the Metaverse uses digital reconstruction and simulation techniques to reverse the research form into a new paradigm: from the mechanism to the phenomenon. In this way, research in the Metaverse can quickly verify hypotheses and conjectures and reveal new laws and principles. For digital life, the Metaverse performs a new kind of digital life by establishing molecular and atomic simulation digital models to explore the origin of life and consciousness, as well as establishing a macro-cosmic digital model to explore unreached science fields such as dark matter, dark energy, and black holes.

The policy layout of the Metaverse: Various countries are all actively deploying the Metaverse. In November 2021, the city government of Seoul, South Korea, announced a Metaverse Platform project to be the first local government to provide public services to citizens.

In July 2021, Japan's Ministry of Economy, Trade, and Industry released a Survey Report on the Future Possibilities and Topics of the Virtual Space Industry, which defined the Metaverse as "a given virtual space in which producers in various fields provide all kinds of services and content to consumers" [4].

In October 2021, US senators from both parties introduced the Government Ownership and Oversight of Artificial Intelligence Data Act. On March 9, 2022, US President Joe Biden signed an Executive Order on Ensuring the Responsible Development of Digital Assets.

The fifth article of the 14th Five-Year Plan of China pointed out that the key industries of the digital economy include artificial intelligence and virtual/augmented reality. The Shanghai 14th Five-Year Plan for developing the electronic information industry points out that it is necessary to strengthen the underlying core technology of the Metaverse, promoting new terminals that deepen perceptual interaction and construct systematic virtual content.

The key technology of the Metaverse: A complete technical system of the Metaverse includes five primary fields: perception, computation, reconstruction, interaction, and collaboration.

The application scenario of the Metaverse: Various scenarios can use the Metaverse:

1. The city-wide digital intelligent system for a city will achieve an entire intelligent IoT perception, a wide range of dynamic responses, and all-around urban security.
2. An interactive collaboration system in the industry will achieve high-precision digital twin, full-chain production management, and full-efficiency industrial collaboration.
3. A whole-system intelligent medical system will achieve interactive teaching on diagnosis and treatment and highly reliable teleoperation.
4. A future national defense system will achieve holographic battlefield planning, compassionate and long-range control, and intelligent and autonomous combat.
5. In scholarly communication, holographic classroom intelligent media will achieve immersive educational learning, cross-temporal information interaction, and visual news dissemination.

3.2 The energy revolution solves climate problems: carbon peaking and carbon neutrality

3.2.1 The energy revolution triggered by the climate problem

3.2.1.1 The nature of the climate problem

The Earth's average temperature has experienced a significant upsurge of 1.2°C, resulting in an array of ecological and environmental issues that pose a threat to sustainable development and human existence [9]. In October 2018, the Intergovernmental Panel on Climate Change released a special global warming report citing 1.5°C. Scientists universally agree that global warming persists and that anthropogenic activities, particularly carbon dioxide emissions, serve as the primary driver of greenhouse gas accumulation. Therefore we should couple the issue of controlling climate change and reducing carbon emissions [10].

3.2.1.2 Big actions on global climate issues

In the face of the increasingly severe climate crisis, the international community turns to both the threat of global climate change for the living space of humankind and the importance and urgency of taking joint response measures to reduce and prevent climate risks. The governance action responding to global climate has already begun. In 1992 the United Nations Framework Convention on Climate Change launched the international system to address climate change. The Kyoto Protocol and the Copenhagen Protocol established the international climate regime from 2013 to 2020. The Paris Agreement officially signed in 2016 consents to control the increase of global average temperature below 2°C above the preindustrialization level and strive to limit it to 1.5°C. Besides, the Paris Agreement promised to achieve carbon neutrality in the second half of this century to reduce the risk of climate change on the Earth's ecology that threatens human development [11].

Major developed and developing countries have successively put forward carbon peaking and carbon neutralization goals. Table 3−1 shows the schedule of each country's carbon peaking and carbon neutralization plans of each country.

3.2.1.3 Energy revolution and climate ambition

The global carbon dioxide emissions reached 33.5 billion tons in 2018, 96% of which were generated by fossil energy. In 2018 the largest source of global carbon emissions was thermal power generation, which emitted 14 billion tons accounting for 42%, followed by transportation emitted 8.3 billion tons accounting for 25%, and industries such as smelting and refining reached 7.8 billion tons, accounting for 23%.

Therefore global carbon emission reduction predominantly needs to rely on reducing the use of fossil energy in the power generation sector, reforming the energy supply, and vigorously developing photovoltaic, wind power, and other renewable energy production. At the same time, to reduce fossil energy consumption, it is necessary to improve the electrification rate in important terminal energy consumption sectors, including transportation, construction, and industry. Electrification, biofuels, and hydrogen energy cannot completely replace fossil fuels for aviation and high-energy-consuming industries. To achieve the above carbon emission reduction goals, we should devote large-scale investment to developing clean and renewable power generation, power grid, and hydrogen production equipment. Transforming energy consumption infrastructure, industrial facilities, and fossil energy power generation is also essential.

Table 3–1 Carbon peaking and carbon neutralization schedule of count.

Country or region	Commitment to nature	Carbon neutral target time	Country or region	Commitment to nature	Carbon neutral target time
Uruguay	Voluntary emission reduction plan under the Paris Agreement	Carbon neutral by 2030	Portugal	Statement of policy	Carbon neutral by 2050
Finland	Ruling party coalition agreement	Carbon neutral by 2035	Switzerland	Statement of policy	Carbon neutral by 2050
Austria	Statement of policy	100% clean electricity by 2030 and climate neutral by 2040	Spain	Draft Law	Carbon neutral in 2050, with a commission to monitor progress on the draft and an immediate ban on new coal, oil, and gas exploration licenses
Iceland	Statement of policy	Carbon neutral by 2040	Hungary	By law	Carbon neutral by 2050
United States California	Executive order	100% renewable electricity by 2045	South Africa	Statement of policy	A net zero economy by 2050
Sweden	By law	Carbon neutral by 2045	Marshall Islands	To submit to the United Nations voluntary emission reduction commitments	Carbon neutral by 2050
Canada	Statement of policy	Carbon neutral by 2050	Korea	Statement of policy	Be carbon neutral by 2050 and end coal finance

(Continued)

Table 3–1 (Continued)

Country or region	Commitment to nature	Carbon neutral target time	Country or region	Commitment to nature	Carbon neutral target time
The European Union	To submit to the United Nations voluntary emission reduction commitments	Carbon neutral by 2050	Bhutan	Voluntary emission reduction plan under the Paris Agreement	It is currently carbon negative and promises to be carbon neutral during development
Denmark	By law	Sales of new petrol and diesel cars will be banned by 2030 and carbon neutral by 2050	New Zealand	By law	Carbon neutral in 2050, when biomethane will be reduced by 24%–47% from 2017 levels
The United Kingdom	By law	Scotland will be carbon-neutral by 2045 and the rest by 2050	Costa Rica	To submit to the United Nations voluntary emission reduction commitments	Carbon neutral by 2050
Ireland	Ruling party coalition agreement	Be carbon neutral by 2050 and cut emissions by 7% a year over the next decade	Chile	Statement of policy	Phasing out coal by 2040 and be coming carbon neutral by 2050
Norway	Statement of Policy (Intention)	Carbon neutral by 2030 through international offsets and carbon neutral domestically by 2050	Fiji	To submit to the United Nations voluntary emission reduction commitments	Carbon neutral by 2050

Country	Commitment	Target
France	By law	Triple the rate of emissions reduction and become carbon neutral by 2050
Slovakia	To submit to the United Nations voluntary emission reduction commitments	Carbon neutral by 2050
Germany	By law	Carbon neutral by 2050
China	Statement of policy	It will peak by 2030 and be carbon neutral by 2060
Japan	Statement of policy	As early as the second half of this century
Singapore	To submit to the United Nations voluntary emission reduction commitments	As early as the second half of this century

These achievements need technical breakthroughs in hydrogen energy preparation, storage and transportation, electric vehicles (EVs), energy storage equipment, offshore wind power, power load response and flexibility, electricity marketization, and bioenergy development [12]. In addition to these changes in technology and hardware facilities, the international community must also carry out iterative updates of soft systems in cooperation mechanisms and goal setting, power market reform, and carbon trading market construction. With the help of such soft systems, the energy system can achieve low-carbon goals through the legal, administrative, market economy, and other regulatory mechanisms and finally reach a high proportion of clean electricity and renewable energy.

Behind the global carbon peak and neutral climate ambitions is an energy revolution involving energy production and consumption, which brings great challenges as well as unprecedented opportunities to related industries.

3.2.2 Opportunities and challenges under the energy revolution

3.2.2.1 The fourth industrial revolution

Digital productivity stands as a defining characteristic of the fourth industrial revolution that distinguishes it from the previous three industrial revolutions and needs to be supported by the economic base and the superstructure. The former relies on three key pillars: energy, information, and transportation networks, while the latter requires the humanity network. By integrating these four networks (i.e., energy, information, transportation, and humanities) and their associated flows (energy, information, material, and value), or 4N4F, individuals' subjective initiatives can fuse with ongoing innovations in energy, information, and transportation. Moreover, the human-information-physical system can pave the way for novel production relationships. Besides, the 4N4F enables tremendous productivity from the data-dividend fourth industrial revolution and exponential growth based on productive forces accumulated during the first three industrial revolutions [13].

The fourth industrial revolution has brought significant challenges and unprecedented opportunities. It is different from the previous three industrial revolutions because it focuses on the digital economy supported by artificial intelligence and forms a new service-centered industry instead of isolated industries, transportation, energy, and information technology. Such changes must combine with the innovative application of emerging technologies, which, in turn, will reshape the productive forces and relations and inject new momentum into the transformation and innovation of the energy industry through digital transformation. The energy industry is experiencing development from a mechanism model to a data drive and now to autonomous intelligence, showing a trend of high integration with the information and transportation industry [14]. Therefore the essence of the energy revolution under the digital economy is combining artificial intelligence and energy technology and interacting with the information and transportation revolution. The revolution will also continue the development process of the digital economy from the e-commerce sharing model to the "Travel as a service" model.

The fourth industrial revolution will fundamentally change the lives of the present and future generations, re-emerging the economic, social, cultural, and environmental conditions on which humanity depends. We need disruptive strategic thinking to deal with the fourth industrial revolution. The integration of 4N4F is an excellent method to support the fourth industrial revolution by replacing traditional linear thinking with comprehensive circular thinking, linking the energy, information, and transportation revolutions, and integrating the supernatant humanity network with the foundational energy, information, and transportation networks. This process then produces subversive economic and ecological benefits (Fig. 3—8).

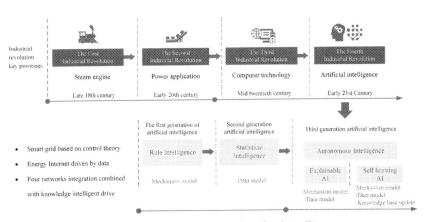

Figure 3—8 The fourth industrial revolution and artificial intelligence.

3.2.2.2 Challenges faced by the energy revolution

With the development of the energy industry toward low-carbon and clean working, the production, transmission, distribution, and consumption of energy are facing profound changes [15]. At the energy consumption end, with the promotion and application of new energy transportation, 5G base stations, and data centers, users' energy demand is diversified and personalized. At the energy production end, fossil energy such as oil and coal, which produce high carbon emissions, has been changing to clean and renewable energy represented by wind or photovoltaic power generation and green hydrogen energy. Although beneficial to the environment, this change process is not totally without a hitch: relevant technologies need improvement, and clean energy policies are conservative. These challenges put forward three requirements for the development of the energy industry:

- Smart energy integration infrastructure needs to be integrated with information facilities to solve the high energy consumption problem using 5G and data centers.
- The energy technology innovation process needs to match the updating of industrial policies to solve cross-industry and joint-innovation problems.
- The business model of the energy industry, under reform, needs constant adaption to the market rules and to address the common service-oriented trend of cross-border integration.

The development fluctuation of the energy industry is not only related to technology but also affected by energy policies as well as industry and market norms. In essence, policies and norms are both born as human subjective initiatives and directly impact industry development, so they are equal to the interaction between human behavior, industry, and technology. Specifically, the interaction is the correlation logic and causal relationship between technological innovation, industrial development of policies and norms, and domain knowledge. Establishing relevant top-level design and theoretical guidance in joint industrial innovation can further strengthen these policies and norms' continuity, consistency, causality, and sustainability. If this interaction process can be digitized, the knowledge and existing data elements can compensate for the present shortcomings of the energy industry development. This interaction is conducive to breaking through the bottleneck of industrial development by establishing the joint innovation of industrial ecology and jointly promoting the energy revolution under the digital economy.

At present, the energy industry is developing from big data intelligence to cognitive intelligence. This development contains a pain point: transitioning from traditional perceptual intelligence to decision-making intelligence. Due to the lack of complete interpretability of big data algorithms, it is difficult for most high-security industrial applications to form a decision loop. This difficulty will greatly limit the deep application of artificial intelligence in the industry. In addition, many industrial policies and technical specifications in the energy industry update behind the pace of technological innovation. Due to the lack of timely traceable correlation logic, the development of the industry has also been affected. This problem has become a bottleneck in industrial and energy Internet applications.

For the energy revolution, it is crucial to understand the new production relationship represented by the digital humanity network and its role in promoting the new digital productive forces. With the in-depth development of the energy revolution, the trend of interaction and integration of humanity, information, energy, and communication networks has gradually manifested, becoming the core content of the transportation and information revolution. The goal of integration comes from combining two into one idea: to obtain a more significant value gain than simple superposition ($1 + 1 > 2$ effect). The integration mechanism comes from system coupling and equipment sharing, and by improving the utilization of energy and resources in the network, the value gains. The subversion is achieved by optimizing the dynamic causality and complex connection logic among the fusion subjects through the digital humanity network. Its economy constantly drives the intelligent dynamic evolution from technology and industry to ecology among the agents through the value stream.

3.2.2.3 Transformation of the electric power system under "carbon peak and carbon neutrality"

The power system is the main link of energy transformation and the key area of carbon emission reduction. Turning down the carbon emissions from power systems is of global significance to achieve the goal of carbon peak and carbon neutral conditions. The transformation from traditional fossil power generation to clean and renewable power generation can reduce carbon emissions and also bring challenges to the power system: how to effectively accept unstable renewable power generation while ensuring security, stability, and efficient operation.

Therefore the key to building a new power system under the carbon neutrality target is planning the future power generation, power grid, and energy storage to meet the load growth in a safe, low-carbon, and economical way. The essential feature of the new power system is to recouple power generation and power consumption through energy storage and a distributed grid. On the one hand, the power supply and demand can be balanced dynamically under the fluctuating renewable energy generation and load through charging and discharging energy storage. In this way, the traditional rigid power system can be transformed into a flexible new one. On the other hand, renewable power generation, power grid, load, and energy storage all couple in a distributed power grid or a microgrid. Through the coordinated and optimized operation of the generation, grid, load, and storage within the distributed power grid, the local consumption of renewable energy is prioritized to reduce large-scale, long-distance transmission and system operation risk, improving the new energy consumption efficiency. At the same time, as a platform for energy interaction, the microgrid can integrate new power sources and loads, such as photovoltaic power generation, distributed energy storage, hydrogen energy, and charging piles. The comprehensive efficiency and reliability of the energy supply can be improved.

The microgrid can transform the power generation mode into a local and decentralized one with clean and renewable energy power generation. Meanwhile, the microgrid should promise the power system stability while connecting nearby power generation and efficiently using renewable energy power. Reducing the construction and operation costs of transmission and distribution networks is also a new challenge. Based on research and operation experience, building a distribution system operation (DSO) center responsible for regional energy dispatching can coordinate the operation of various factors in the distribution network. Furthermore, the DSO will also support the carbon securities and energy trading center based on blockchain technology. The DSO responsible for regional energy dispatching includes applications such as microgrid analysis, calculation, coordination, low-voltage user power monitoring, and commercial insurance services. These functions can effectively control the consumption, distribution, and storage of distributed energy to obtain the maximum value from the microgrid.

For example, a typical microgrid application close to the user side is the integration of optical storage and charging. The microgrid acts as both an energy producer and a consumer. In the microgrid, the DC and digitalization of low voltage on the terminal power side can realize the energy flow dispatching management. Unified intelligent management of distribution and charging networks and intelligent collaboration of optical charging and discharging enable EVs to be more efficient and have lower carbon emissions. The external microgrid interacts with other microgrids or upper energy networks in multiple streams to transfer energy supply margin and demand space information and carry out the migration of microgrid energy in time and space. Based on AI's continuous learning of market and load, it can accurately judge future energy use and actively participate in frequency modulation and peak shaving. This microgrid has also become a link between intelligent transportation and smart energy. The mobile energy storage carried by tens of millions of EVs will intelligently integrate into the 4N4F. The energy flow will become bidirectional, showing rich V2X scenarios such as V2G, V2H, V2V, and V2B. At the same time, it can consider the integration of communication base stations and data centers so that electricity will enter from the network, expanding the microgrid's regional scope and application scenarios and iteratively upgrading to serve more industries and individuals (Fig. 3−9).

3.2.3 Development of energy revolution under the 4N4F

3.2.3.1 The future of the energy revolution

The 4N4F represents the fundamental changes and influences of the fourth industrial revolution on the philosophical, scientific, and engineering levels of human thinking. Furthermore, it advances the industrial

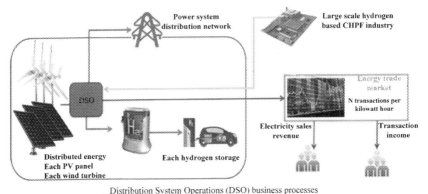

Figure 3−9 Distributed network planning of energy ecosystem.

strategies required to incorporate philosophical thought, scientific theory, and engineering practice.

From the philosophical perspective, it is necessary to establish a change from simple linear thinking to global circular thinking between the economic base and the superstructure and form a value-added effect that the whole is greater than the sum of the individuals. The coupling of energy flow and material flow can realize value transformation, which is the basic attribute of sustainable ecosystem development. In addition, the coupling reflects the philosophy that the whole is greater than the parts. The local conversion of individual energy flows and material flows will reduce their own value, but if the two couple together, the overall value can be improved. The specific solutions and benefits [16] of the above coupling are as follows.

- By repeatedly iterating the energy and substance's self-similarity value conversion process, sustainable circular ecology can be achieved from three dimensions: time, space, and type, equivalent to using energy/ material coupling to solve energy storage problems.
- Sustainable circular ecology can continuously improve energy intelligence by transforming useless resources into useful ones. In this process, the local value declines but the overall value increases.
- Networks like microgrids, distribution networks, and regional energy networks are self-similar: the mobile transportation network is similar to the animal community, and the low-carbon building network is partly the same as the plant community. Based on this, the 4N4F system method can simulate the self-organizing fractal structure of the natural ecosystem. Through the exchange of energy and materials with each other, continuous iterating and evolutionary progression will form an urban smart energy ecosystem (Fig. 3−10).

From a scientific perspective, the 4N4F reveals the internal law and the basic interaction relationship between energy, information, and human behavior, turning waste energy into useful and promoting carbon neutrality. Like Einstein's revelation of the identity of mass and energy through the mass-energy equation, Prof. C. C. Chan, in 2013, proposed and developed the idea of establishing a smart energy development path by studying the internal correlation between energy and information. He pointed out that human interaction behavior (the humanity network) is the key to intelligently blending energy and information carriers. Its typical scenario is intelligent transportation with the Internet of Vehicles

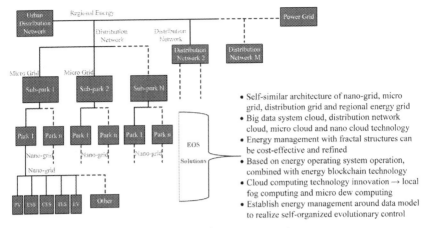

Figure 3—10 Self-similar characteristics of energy network system.

(IoV) and automatic driving powered by electrical energy and hydrogen fuel. The challenges and critical solutions are as follows:

- To realize the value transformation of energy flow and material flow, two challenges need special attention: one is the limit of entropy increase by the second law of thermodynamics, and the other is how to ensure the system's sustainability through continuous increments.
- The open characteristics of energy flow and material flow can be established by coupling the information flow, which can transform the closed local systems into open and integrated systems. In this way, the useless energy will become useful, and engineering will be realized through value communication.
- The value gradient of each link of industrial ecology is the key to driving the coupling of energy flow and material flow, whose value transformation always points toward increasing the value gradient (Fig. 3—11).

From the engineering level, we will combine energy technology with information technology to establish an intelligent energy operating system platform with an artificial intelligence/big data system. Additionally, we will adopt an end-to-cloud architecture to integrate the energy and information grid and obtain the maximum value-added benefit of industrial collaboration.

- The smart energy operating system is equivalent to the endogenous basic framework and rules for the sustainable development of the ecosystem. It is also a platform for the intelligent integration of humans and machines.

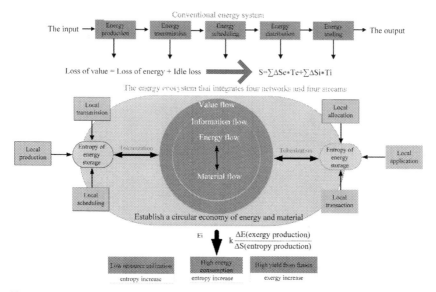

Figure 3–11 Energy, material, and value stream of energy ecosystem.

- Due to the self-similar structure and characteristics of the intelligent energy operating system, the basic implementation of human—machine intelligent fusion can occur on the edge of the cloud side.
- The intelligent energy operating system is the digital mapping and carrier of the "human-information-physical" integration and is the engine for realizing smart energy.

Under the 4N4F, the basic characteristics of energy system development are:

1. Based on the integration of energy flow and information flow, the digital characteristics of data empowerment and knowledge drive are formed:
   a. Energy big data platforms have been extensively built and widely integrated with cloud computing, but they still lack an effective means of data-to-knowledge conversion.
   b. Traditional energy devices are upgrading edge computing to enable IoT access, but they are still facing the challenge of increasing energy consumption in data centers.
   c. 5G networks have begun to be vigorously developed to fully promote the application of artificial intelligence. However, they still face the challenge of high customization and the high energy consumption of networks.

2. Based on the integration of energy and material flow, the cross-industry characteristics driven by low-carbon travel are formed:
   a. From fossil fuel vehicles to EVs is a cutting-edge application in the energy field that integrates transportation electrification, but it still faces multiple challenges in battery technology.
   b. From EVs to fuel cell vehicles is a cutting-edge application of energy fusion involving hydrogen transportation, but it still faces the multiple challenges of fuel cells.
   c. Charging/hydrogenation networks are an important integrated infrastructure for decarbonizing energy and transportation. However, they still face multiple challenges.
3. Based on the integration of value streams, the service characteristics driven by user behavior are formed:
   a. The main body of the power generation side can optimize the joint operation of multiple energy assets through the virtual power plant model. However, it still faces challenges in industrial policy.
   b. Demand-side users can optimize the cost-effective operation method of energy terminals through demand-side response but still face user engagement challenges.
   c. End-side energy managers can form regional energy optimization solutions through microgrid construction, but they still face various challenges in economic operation.
   d. Energy system participants can autonomously form energy transactions through the trusted energy blockchain network but still face multiple security challenges.

The above characteristics show that the transformation and development of the energy industry can no longer only rely on technological innovation within the energy enterprise to make breakthroughs. It must rely on the collaborative innovation of the industrial ecology to establish a strategy adapted to the integration and development of the industry from a new perspective. Driven by the 4N4F, the integration innovation formed between industries will become a new industrial ecology, which has been fully reflected in energy storage and new energy vehicles. By doing this, the intelligent integration of multiple types of systems caused by the four major energy contributions can be solved at a higher cost to meet the personalized functional requirements, resulting in a low return on investment of the system and the inability to obtain effective applications.

3.2.4 Typical cases of 4N4F in the energy revolution

The industrial strategy of 4N4F integration can break the barriers between various links inside and outside the industry, promoting the intelligence of energy ecosystems from a top-level design perspective, with the ultimate objective of transforming the energy industry into a service-oriented one. A practical example is the optimized operation of photovoltaic-storage-hydrogen-charging at the terminal side, where we can achieve the zero-carbon energy-ecological operation mode of transportation/building coupling in smart cities.

Based on the idea of 4N4F, the Beijing Low Carbon Clean Energy Research Institute demonstrated and verified a smart energy system in a park by establishing a self-similar energy network structure integrated with edge computing, a nanogrid research platform, and a microgrid demonstration project. The demonstration project combines the localized operation of energy services with the networked management of energy systems (NicerNet) based on the operating system approach (NICE_EOS). Through the virtual private cloud (NICE_CLOUD), edge computing (NICE_FOG), and end gateway (NICE_MIST) architecture, it realizes the integrated scheduling and optimal utilization of multiple energy sources such as light, storage, hydrogen, and charging in the park (Fig. 3—12).

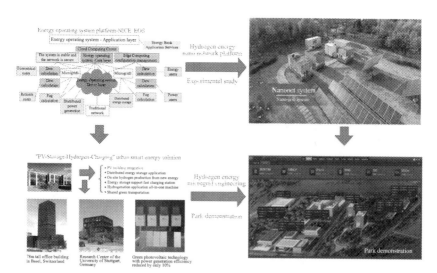

Figure 3—12 Distributed operation design of energy ecosystem.

Finally, through the NICE_EOS and NicerNet, a knowledge base platform for the energy industry can be established, directly integrated with optimizing self-similar district energy system operations. Based on the ecological energy map, the evolution process of data-information-knowledge-intelligence is established through interaction with the system operation data. A carbon balance (neutral) energy ecosystem is formed (Fig. 3–13).

3.3 Changes in mobility patterns brought about by the transportation revolution

3.3.1 Transportation reform in the contemporary era

The transportation industry refers to the social production sector in the national economy that specializes in transporting goods and passengers—for example, highways, railways, aviation, and water transportation. With the new round of scientific and technological revolution, countries have accelerated the transformation in the field of transportation.

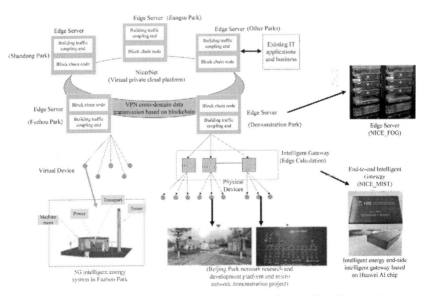

Figure 3–13 Distributed operation demonstration of a carbon-balanced energy ecosystem.

On the one hand, with the deepening of the structural reform in the energy supply side, traditional energy, such as coal, has declined in production. Countries have strengthened renewable energy applications in industries, among which automobile manufacturing is the most coupled with the new energy industry [17]. In addition, with the strong advocacy of the global green concept, the low-carbon fuel used in aviation and shipping transportation has gradually received attention.

On the other hand, the rapid development and application of various new technologies have made intelligent transportation a reality [18]. Intelligent transportation means making full use of the Internet, space perception, cloud computing, mobile Internet, and blockchain to produce a new generation of information technology. Intelligent transportation also comprehensively utilizes transportation science and system methods such as artificial intelligence theory and tools to achieve comprehensive perception, depth fusion, active service, and scientific decision-making as the goal. As a result, it can achieve resource optimization for the public service industry and promote safer, more efficient, convenient, economical, environmentally friendly, and comfortable transport industry.

With the development of urbanization, urban transportation reform is imminent. Under the concept of smart transportation and green travel, "Travel as a Service" is proposed. "Travel as a Service" is a kind of people-oriented business model and travel culture. Its essence is to integrate multiple vehicles and travel patterns based on a unified service platform. The core of the "Travel as a Service" system is the traveler, and the principle of the system is data-sharing service using big data technology to optimize the allocation of resources and establish a seamless decision. Also, using new mobile payment methods to provide more flexible, efficient, and economical travel services that meet the needs of travelers. The realization of "Travel as a Service" requires the integration of four aspects: travel demand (traffic users), travel carrier (manufacturers), travel route (government regulations), and travel conditions (operating environment). At present, it is reflected inside transportation carriers, such as the IoV (human-vehicle-road-environment), smart subway, high-speed rail network, and smart airport. In the future, it is necessary to realize the connection and integration among different transportation carriers to provide the most personalized intelligent transportation services while meeting the different needs of industrial chain participants. The emergence of this intelligent transportation mode will fundamentally change the product definition and value chain system of vehicles and other transportation carriers (Fig. 3—14).

3.3.2 Changes in the transportation industry brought about by the 4N4F

3.3.2.1 Digital transformation of the transportation industry

Taking the IoV as an example, Fig. 3—15 shows that the car has changed from a mechanical product to a software-defined digital product in the digital economy. In the development stage of the passenger economy, the car will further develop into an accessory to the user terminal service. Therefore original equipment manufacturers (OEMs) need to adapt to the transformation of traditional models and engines to digital products from now on. More importantly, in the process of industrial chain integration, they should understand again the real service demand behind digital products [19]. This change in demand is mainly reflected in three aspects:

1. Technology strategy and product positioning: the single technology development model has been transformed into a multifunctional platform. For example, algorithms and cloud computing process a large amount of data collected by cameras and lidar in the field of autonomous driving to carry out a variety of simulation analyses and realize various functions. Based on the cloud edge collaborative mode, the product performance of vehicles can be upgraded iteratively, and the function evolution can be evolved through remote software updates. Cars have become software-defined upgraded products, and this mode has been used for many years on other transportation carriers such as airplanes.

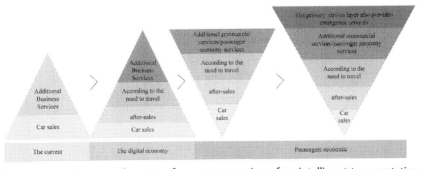

Figure 3—14 The core elements of a new generation of an intelligent transportation system.

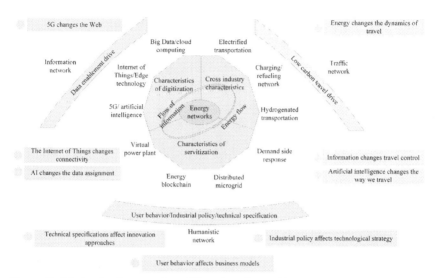

Figure 3–15 Application practice of Shenzhen Smart Airport.

Example 3.1. Tesla intelligent system.

With energy technology development, the automobile revolution has been both electric and intelligent. Electrification only changes the power supply mode of the car, while intelligence is the core of the subversive changes to the car. It will transform the car from a traditional mechanical body into an intelligent body with powerful computing power [20].

In this revolution, Tesla plays an irreplaceable leading role. Tesla is the only company in the world that has realized full-stack self-research and self-production in the core field of automatic driving. It has built a set of software and hardware architecture of full-link automatic driving, including sensing, control, and execution in data, algorithm, and computing power levels.

Tesla electric cars are with pure visual perception: completely devoid of sensors such as laser or millimeter wave radar. These cars only perceive through eight cameras capturing the real-world image data and process these data through the complex neural network perception architecture to construct 3D vector space of the real world. In vector space, a hybrid planning system combines traditional control methods and neural networks to realize the vehicle's behavior and path planning. Then, a control model is generated and transmitted to the actuators. Meanwhile, the complete data closed-loop system and simulation platform realize the continuous iteration of autonomous driving capability.

In order to improve the safety of automatic driving, Tesla adopts a large number of intelligent simulations and extreme scenes to build a virtual simulation space in the real world. These simulations and scenes are difficult to encounter in the real world, and some complex ones are quickly annotated. Up to now, the scale of virtual data obtained by Tesla through simulation has reached 3.71 billion pictures and 480 million annotations, which have been integrated into the vehicle model. These data will greatly train the algorithm's performance to cope with different scenarios.

In addition, Tesla is also using "over-the-air" (OTA) download technology to realize remote software management through the interface of mobile communication. At present, upgrading software through OTA is widely used in smartphones.

OTA upgrades for cars are like Windows upgrades for computers or mobile phones. Each upgrade can lead to improvements, bug fixes, more features, performance improvements, or visual improvements. This kind of update should pass the following procedures: online network detection, matching version, new code downloading to the local user, installation execution, and verification.

Tesla's Model S, launched in 2012, and the latest Model 3 have full OTA capability. OTA upgrades are not only available for in-vehicle entertainment systems and applications but also for software updates to the ECU, such as the battery management system, the electric drive control unit, and the vehicle control unit.

Tesla has acquired enough operation data through large-scale automobile production. These data continuously train the iterative full self-drive (FSD) algorithm and supercalculate force service AI to realize better performance. Ulteriorly, better algorithms promote a whole vehicle intelligent system with remote management and online upgrade installation, finally helping achieve fully automated driving and making the software-defined car a reality.

2. Industrial synergy and user demand level. Industrial manufacture and product selling will gradually transit from the simple linear industrial chain to multiparty cooperation and user participation as the core of operator mode. New mobile Internet technology will be applied throughout all the stages, including marketing, sales, and the user

experience, to provide better products, services, and personal value more efficiently and conveniently. Following users' online full-path mobile Internet consumption, namely the purchase (flow rate), pay (value stream), service (energy flow), and offline experience (material) stages, data can be integrated to complete the user-centered precision product marketing and sales. In this way, the above 4N4F can be used to provide customers with the most convenient service.

Example 3.2. Mercedes—From Serving "Cars" to Serving "People."

The traditional sense of service is for cars, such as maintenance and repair. Nowadays, the emerging service concept pays attention to the service for people, and the car has become just a service carrier. This change is the inevitable result of the market's continuous maturation, modernization, and interconnection, and it is also the focus of the future competition of major automobile enterprises.

Based on the Internet, Mercedes-Benz provides users with more types and more efficient life experiences. During COVID-19, car companies launched a "no contact" service. They simplified the online process and constructed a digital system to improve customer experience. In addition, they also provided users with more convenient services such as online shopping, a tire key refresh, and online after-booking. They started the car direct-selling business in Sweden and South Africa.

In terms of online, with the gradual rollout of 5G in China, Mercedes-Benz has built four online digital platforms for users: the website, WeChat, application, and car machine with online and offline connection services. In addition, Mercedes-Benz launched the "Mercedes Benz EQC online Showroom" mini program in June this year, providing users with rich functions and more interactive experiences such as 3D car viewing and test drive reservation. In special periods, Mercedes-Benz makes short personalized videos for customers to reduce contact and save time. These videos are more intuitive and personalized to provide relevant information to customers instead of traditional telephone and E-mail forms.

On the offline side, Mercedes-Benz has innovated three aspects: in terms of space and specifications of retail outlets, it develops user-oriented architectural design styles. In multimedia and digitalization, it transforms into innovative digital service hubs for users. In terms of process, it realizes the multidimensional network of sales and service processes for users with tools such as Benz App. Meanwhile, it provides rich digital services in the retail points of automobile enterprises, for example, the digital detector project, which is currently under exploitation. When the vehicle passes, the digital detector will read the plate, measure the tread depth, and photograph all angles, including the chassis, to record the vehicle in detail and inform the maintenance consultant of any visible damage. The information will then be presented to the customer clearly and concisely.

With the release of the "Best Customer Experience 4.0" strategy, Mercedes-Benz intends to launch innovative platforms such as Her Mercedes, Mercedes me Club, and "Number one Territory," specially created for the new generation of customers. These platforms can complete the precision product marketing and sales with users as the center to grasp the 4N4F and provide customers with the most convenient service.

3. Business model and lifestyle: Under the pile-network cooperation model, charging piles can form an appointment with the EV's battery management system, reducing car owners' waiting time and optimizing batteries' health by charging on demand. Based on cloud edge collaboration, users can live and travel more freely with improved charging convenience and reduced mileage anxiety. As mobile energy storage, EVs can also interact with smart parks and microgrids and actively participate in the business process of energy flow [21].

Example 3.3. Vehicle-To-Grid (V2G) Experimental Project in Turin, Italy.

One of the challenges in pile-network collaboration is the interaction between EVs and the grid. The V2G will solve this challenge [22].

The core idea of V2G is to use energy storage on vast EVs as the buffer between the grid and renewable energy sources. EVs feed the stored power to the grid when the load is too high. When the load is less than power generation, EVs store the excess power to avoid waste. In this way, EV users can buy electricity from the grid when the price is low and sell electricity to the grid when the price is high, thus obtaining certain benefits.

Since September 2019, Fiat Chrysler Automobiles (FCA) has selected ENGIE Eps as the technical partner for the world's largest V2G pilot project and collaborated with Terna to cotest EV feedback. Later, FCA announced that COVID-19 had not stopped the project from partnering with ENGIE Eps and Terna. The project's first phase began at the FCA plant in Mirafiori, Italy, aiming to enable two-way interaction between FCA BEVs and the grid. The construction site is now open at the Drosso Logistics Center in the Mirafiori building, covering an area of approximately 3000 square meters. A 450 m trench has been excavated to accommodate more than 10 km of cable, which connects the grid to 64 two-way fast charging points with a power output of up to 50 kW. Engie EPS designed, patented, and built the centralized infrastructure and advanced control systems to provide vehicle-to-grid networking services in addition to fast charging of EVs.

By the end of 2021, the charging infrastructure can be expanded to install two-way charging systems for up to 700 EVs, providing a regulatory capacity of up to 25 MW, making it the largest ever V2G facility globally. At the same time, along with other FCA assets, including 5 MW solar panel capacity, FCA will build a massive virtual power plant providing optimized power for 8500 homes and operation suggestions, including ultrafast frequency regulation.

3.3.2.2 Cross-industry transformation of the transportation industry

Based on the integration of energy flow and material flow, the cross-industry characteristics of low-carbon travel drive can be formed. At the same time, the integration also meets some challenges.

1. The frontier integrated transportation electrification in the energy field attributes the evolution from fuel vehicles to EVs. Still, it faces various challenges, like battery technology.
2. Hydrogenation technology lets EVs evolve into fuel cell vehicles; however, the technology is not yet mature.
3. A charging/hydrogenation network is important for reducing carbon emissions in energy and transportation, but it still faces multiple challenges.

Once these challenges are overcome, the industrial strategy of the 4N4F can function better:

1. breaking down the barriers between various links inside and outside the industry;
2. promoting the intelligence of the energy ecosystem from the top-level design;
3. realizing the transformation from industry to service.

For example, relying on the 4N4F optimization operation of "hydrogen storage and charging" at the terminal side, the smart city's zero-carbon ecological operation mode of transportation/building coupling can be realized (Figs. 3−16 and 3−17).

3.3.3 Specific changes brought about by the change in transport pattern

3.3.3.1 The transportation system is highly interconnected

The 5G IoV should achieve high cooperation in five dimensions [23]: people, vehicles, road, network, and cloud. Multiple means of travel or logistics transportation solutions need an optimized door-to-door service path to integrate travel or logistics needs, multimode transportation, information, and energy networks. For travel services aimed at people, the core is the one-stop solution where consumers become intelligent travel terminals. For transportation carriers, the future transportation tool combines the data sender and receiver and the computing and data-sharing node and will become a distributed data center. It will use communication modes to integrate the collected and released traffic information from intelligent roadside facilities and use local edge computing capability for

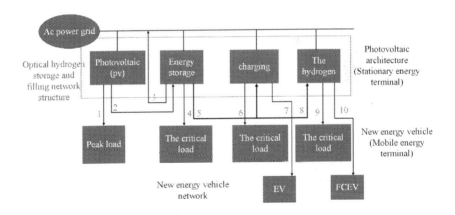

1- photovoltaic peak cutting, 2- optical storage integration, 3- energy storage peak cutting. 4- energy storage standby, 5- energy storage charging 6- electric vehicle for power grid, 7- photovoltaic power generation, 8- energy storage for hydrogen production, 9- hydrogen energy reserve, 10- hydrogenation realization

Figure 3−16 The basic characteristics of 4N4F integration.

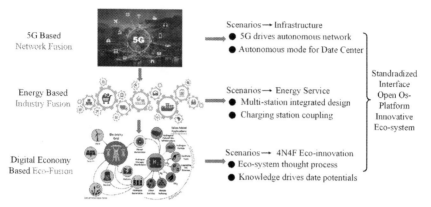

Figure 3–17 4N4F based on optical hydrogen storage.

traffic paths. They are all provided in a smart mobile IoT platform. In terms of network, the two core capabilities of the 5G network, mobile edge computing and network slicing, will build a flexible mobile IoT network. In terms of the cloud, it will build an integrated open data public service platform and cloud control platform and form a powerful cloud service infrastructure through cloud edge collaboration.

Example 3.4. Huawei and Shenzhen Airport.

In order to achieve the goal of "making people happy and making things better," Huawei has launched the Transportation Agent project for the whole business process, including passengers, goods, and vehicles. By integrating connection, cloud, AI, ICT, and data fusion technologies, the project precipitates industry assets and realizes transportation business fusion in four flows: value, energy, traffic, and information, improving traffic safety, efficiency, and passenger service experience.

In 2017 Huawei and Shenzhen Airport strategically cooperated to promote the airport's digital transformation. After communicating key business processes, they effectively improved the operation efficiency and the service experience. Key technologies are as follows.

1. The integration technology of end-to-end cloud collaboration. It digitalizes big data, video, and AI, constructs a data-information-knowledge-intelligence system through data assets, and unifies the architecture and platform. Sixty business rules digitization tests have been conducted.

2. Distribution power system. This incorporates multiple IT systems and OT operation data by reusing large data modeling and AI machine learning technology.

3. The "variable taxiing time" prediction algorithm. It uses big data and the AI algorithm to improve the situation awareness and coordination of complex scenes through accurate prediction.

These technologies greatly enhance operational efficiency and save time. The artificial work needs 4 hours before it can be automatically assigned for 1 minute. The flight velocity and range increase by 5%. More than 90 flights a day can stop at the bridge, which means 2.5 million passengers can no longer take the ferry. The ground taxiing time decreases by 1−2 minutes, saving 20 million aviation fuel energy annually.

The development of 5G technology has brought about the integration of network, industry, and ecology to the overall development of the digital economy from three levels [24] (Fig. 3−18).

First, 5G communication and data center infrastructure promotes information construction and network integration. The Shenzhen Smart Airport, therefore, put forward an automated driving mode based on the autopilot network data center. According to the different levels of man-machine alignment, the airport defines the self-organizing intelligent path in the future network and data center. Among them, although the autonomous driving mode originated from traffic, it has the important significance of empowering the whole industry [25].

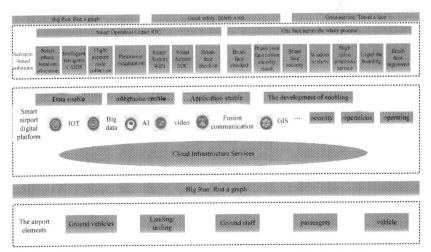

Figure 3−18 Network integration, industrial integration, and ecological integration.

After more than 3 years of exploration and practice, Shenzhen Airport has achieved phased digital transformation progress, effectively guaranteeing the rapid growth of its international and domestic business. In 2019 the growth rate of domestic and international passenger volumes ranked second and first, respectively, among large airports in China. As the benchmark of Shenzhen's pioneering demonstration construction, Shenzhen Airport will further explore the 4N4F with the new infrastructure as the traction and achieve the following goals: a warmer hub for passengers, a more efficient portal for the city, and a pioneering demonstration sample for the industry. The critical practice in Shenzhen will be promoted to the whole industry and become the core solution of "four-type airport" construction by the Civil Aviation Administration.

3.3.3.2 The transportation system is highly automated and intelligent

Intelligent transportation can achieve continuous evolution by comprehensively solving problems [26] with people, vehicles, and the environment on the roadside, taking the IoV as an example. A 5G IoV operating system is necessary for independent, intelligent transportation resulting from multiparty cooperation in the industrial chain. The 4N4F, based on the knowledge graph, can further facilitate the upgrade and evolution of intelligent transportation. As shown in Fig. 3—19, the Travel as a Service model reflects the trend of multiindustry integration and becomes a typical application scenario for the 4N4F. The five-dimensional coordinated development of "people-vehicle-road-net-cloud" will enable the 5G IoV to explore personal services, industrial services, and public management services. Huawei's application of digital transformation in Shenzhen Airport has verified its enabling effect. The autopilot levels are shown in Table 3—2.

The data center is also a typical application scenario. In the traditional mode, energy flow and information flow are independent. After multiple flows integrate with the information flow, the total energy flow receives an optimization strategy from a global perspective. It conducts real-time examinations for its link equipment to achieve a more reliable operation guarantee. Energy saving and emission reduction are the most important evaluation indices during data center's operation. Through AI and other information flow technologies, the operation information of the data center is collected and analyzed, dynamic modeling is performed, and the optimal cooling and power supply strategies are deduced in real-time. At the same time, the overall energy efficiency of the data center is optimized

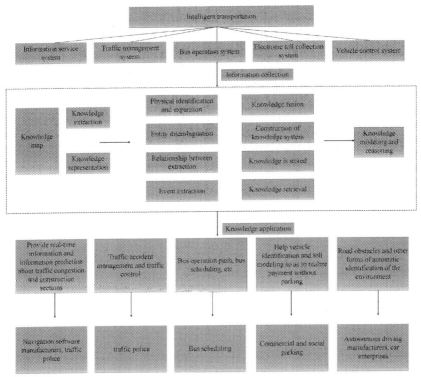

Figure 3–19 Knowledge graph drives the evolution of the Internet of Vehicles operating system to intelligent transportation.

by combining the computing power demand of the upper data stream. In the daily operation and maintenance process of data centers, AI can also be used to carry out predictive maintenance of key equipment in combination with operation data and finally achieve the "autonomous driving" of data centers.

3.4 People oriented for four networks for fusion

3.4.1 People-oriented Industry 5.0

3.4.1.1 The origin of Industry 5.0

Industry 1.0 was the age of machine making, introduced in the 18th century and developed from the 1760s to the mid-1800s. Industry 2.0 is the

Table 3–2 Autopilot mode in the data center.
Data centers on autopilot

Operations	Operating	Autopilot level
Totally manual, manual recording	Completely without optimization	L0
Electronic and digital operation and maintenance data, reduce manual inspection content	Energy efficiency basic monitoring and visualization	L1
Using AI technology to identity the status of dumb devices eliminates routine inspection	Rule-based PUE optimization	L2
AI technology is used to intelligently rectify the problems in the process of equipment operation, eliminating manual analysis	Automatic optimization of PUE energy efficiency based on AI, with manual intervention	L3
Use AI technology to predict equipment health status and preventive maintenance	Energy efficiency optimization based on reinforcement learning; business forecast	L4
Fully autonomous operation, self-repair system failure	Optimal operation is achieved automatically without intervention	L5

age of electrification and automation, flourishing from the second half of the 19th century to the beginning of the 20th century. Industry 3.0 is the electronic information age that began in the 1970s and has continued. Germany put forward the concept of Industry 4.0, which explains how industries, enterprises, and even countries use information technology in a new generation to present a new production mode. Specifically, Industry 4.0 uses digitalization to break through the five links of product design, planning, engineering, implementation, and service [27].

Since its inception, Industry 4.0 has reinforced the highly transformative impact of digital, data-driven, and connected industries. However, Industry 4.0 does not emphasize the importance of industry in providing long-term services to humankind globally. Additionally, it has been noted that Industry 4.0 inadequately addresses the question of how technological innovation can be leveraged to foster collaboration and mutually beneficial interactions between industry and society. By contrast, Industry 5.0 is a human–centered and knowledge-driven era of wisdom. Industry 5.0 systematically proposes to put labor at the heart of industrial production, thereby achieving social objectives beyond employment and growth and providing robust prosperity.

3.4.1.2 Definition and characteristics of Industry 5.0

Industry 5.0 evolved upon the four existing industrial revolutions to different degrees. The European Commission has integrated the concepts of human-centered, sustainable development, and resilience into the development of European industry during Industry 4.0. The supplemented and expanded version of Industry 4.0 is called Industry 5.0 [28], whose definition was given by the EU: Industry 5.0 recognizes the power of industry, that is, the industry production respects the boundaries of the planet and the interests of industrial workers are put at the heart of the production process. In this way, social objectives beyond jobs and growth and robust prosperity can both be achieved. The first white paper on Industry 5.0 released by the European Union mentioned that Industry 5.0 has three characteristics: human-centered, sustainability, and resilience [29].

1. Human-centered: Industry 5.0 will pay attention to social problems, return to the human-centered value positioning, and serve every individual in society. In the past, the industrial revolution mainly replaced heavy manual labor, dangerous production activities, or part of mental work with automated equipment, artificial intelligence, and industrial software. In this way, the industry could improve production efficiency, safety, and consistency of product quality and liberate humans to do more innovative work. However, when such automation and information technology represented by artificial intelligence, the IoT, cloud computing, big data, and mobile Internet reached a new generation, they also brought many social and even ethical problems. For example, will artificial intelligence gradually replace or even exceed human beings? Will the dominance of humans be completely replaced by machines? Will more people lose their jobs? Industry 5.0 provides a clear answer to these questions by returning to a people-centric value proposition. Specifically, we should establish the human beings' dominant and decision-making positions in manufacturing processes and industrial systems. At the same time, we should pay attention to human safety, comfort, employment, training, and skill upgrading.

2. Sustainability: Industry 5.0 will pay attention to environmental and energy issues and sustainable industrial development. The previous industrial revolution paid more attention to the effects that productivity improvement and production mode change have on the social economy. With the gradual decline of nonrenewable energy sources and the frequent occurrence of extreme weather events due to global warming, industrial development must pay attention to sustainable and

green manufacturing. In addition to Europe, China has also put forward the long-term goal of "carbon peaking and neutrality."

3. Resilience: Before the COVID-19 pandemic, the world paid little attention to the resilience of industrial systems, so the industry has been somewhat affected. Industry 5.0 takes the warning and will focus on two aspects of resilience to cope with global emergencies: the first is the company's ability to cope with external uncertainty factors, such as market, supply chain, and customers, and the second is a broader range of the industrial system, such as the whole production chain and even a country or region's resistance ability to deal with unknown risks, such as global emergency, or technology blockade.

3.4.1.3 Industry 5.0 and the Metaverse

The human-centered characteristic in Industry 5.0 will extend from producers to consumers to realize the integration of production and consumption [30].

1. Through intelligent man-machine cooperation, Industry 5.0 no longer requires the physical strength of workers so that the old and weak labor force can return to the factory, which can alleviate the problem of labor force population reduction caused by population aging.

2. Industry 5.0 will combine with advanced virtual—reality interaction equipment, for example, a variety of sensors used in the Metaverse, to design and produce with the help of data interconnection and interflow. Users can get personalized products and improved experiences, while producers can receive real-time feedback as sources of innovation.

3. The cooperation between workers and robots must still be carried out in a shared physical space. With the advent of the Metaverse, people can produce and process products anywhere outside the working space of cooperative robots: advanced sensing equipment obtains the workers' intentions, then collaborative robots will quickly pick up on such intentions and perform specific tasks. The workers can accurately control the physical space where the cooperative robot lies through advanced virtual—real interaction equipment. In the future, the work related to manufacturing will be done in a world parallel with reality but able to affect it. A real connection between the virtual and the real world will be achieved.

Using interactive fusion technologies like digital twin, AR/VR/MR, and brain-machine interface, the enterprise packages itself as a virtual factory in the Metaverse, whose data synchronizes with reality. In this way, factors and processes can be visual, analyzable, intelligent, and precisely controlled. This method can significantly reduce operation, maintenance, production cost, improve production efficiency, and achieve green sustainability and intelligent manufacturing. As a result, the Metaverse can realize 3D immersive mapping parallel to the real world, real-time optimize the process of intelligent manufacturing, and achieve broader synchronous and iterative optimization. In addition, it can also find potential problems and validate the design, manufacture, and processing problems of the production line and equipment structure. At the same time, consumers can participate in all aspects of product design and production at any time through advanced virtual—real interactive equipment such as AR/VR/MR and brain-computer interface. They can realize innovation between production and consumption and provide all participants and stakeholders in the manufacturing value chain with more experience, interaction, and humanity.

3.4.2 Digital humanity network and its application in 4N4F

As the essence of digital, the humanity network is the new production relationship between human-cyber-physical systems under the digital economy. Apart from the big data based on user behavior, human-oriented cognitive intelligence, such as business models, energy policies, industry regulations, and market rules, is also essential for digital humanity network. For example, digital humanities in energy industry ecology are the interaction and rules between human behavior, information system, and the physical system composed of energy and transportation. These factors reflect the fusion of knowledge and data that is necessary for the intelligentization of an intelligent energy system.

The digital humanity network plays an essential role in the industry:

For transitioning energy industry, the digital humanity network is the key link that connects the energy, information, and transportation network and drives the intelligent coupling of energy and material flow through the value flow. The intellectual depth of such coupling will also iterate, upgrade, and evolve through the information flow and form a 4N4F integrated industrial ecology.

For industry innovation, the humanity network can directly interconnect upstream and downstream industries and form an industrial knowledge base to provide necessary ideas, methods, models, and platforms for joint industrial innovation.

For the business model, the humanity network integrates other networks to bring new service types and user experiences which meet individual needs by studying user behaviors and habits. Through the new market elements generated by those customers, the larger-scale market growth can be formed.

For industrial development, the humanity network is the superstructure, whereas the energy, transportation, and information networks are the economic foundation. The integration of the four networks reflects the relation between the new production (digital humanities) and new productivity (energy, transportation, and information) under the digital economy. The 4N4F interaction can better promote and release the vast potential of digital productivity.

The 4N4F with Industrial Humanity Network is shown in Fig. 3—20. The 4N4F plays an essential role in the industry:

Through the 4N4F, the technical specifications, domain knowledge, safety specifications, and management standards can be automatically transformed into a knowledge spectrum and base through semantic models. The base collects knowledge intelligence, such as expert knowledge and industry standard, and integrates with machine intelligence based on big data. Its significance lies in the technical framework of human—machine hybrid intelligence, which allows breakthroughs in explainable artificial intelligence and points out the direction for establishing autonomous intelligence through converting data to knowledge.

Through the 4N4F, the industry has gradually transformed from a linear industrial chain that simply manufactures and sells products to a multi-collaborative operator model that focuses on users. With the help of the new mobile Internet technology, the industry runs through marketing, sales, and users to provide better products and services for users. The 4N4F focuses on the consumption path in mobile devices and collects complete path data, including the purchasing phase (information flow),

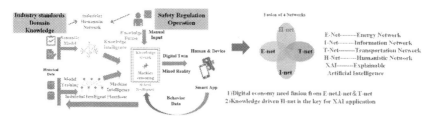

Figure 3—20 4N4F with Industrial Humanity Network.

payment phase (value flow), service phase (energy flow), and offline experience phase (material flow). In this way, we can achieve a convenient user-centric (humanity network) service and precise product marketing and sales system.

The 4N4F reveals the internal interaction mechanism and the primary relationship between energy, information, and human behavior. Prof. Qingquan Chan proposed and developed the idea of establishing a smart energy development path [15] by studying the internal correlation between energy and information [14] in 2013. He also pointed out that human interaction (humanity network) is the key for energy and information carrier to become an intelligent fusion. The typical applied scenario of this idea is intelligent transportation, including IoV and autonomous driving power by electricity and hydrogen. The interaction between productive forces and relations under the digital economy is manifested in the integration of people-information-energy-transportation in the energy system. Among them, the intelligence of energy and transportation represents a new type of productive force, and integrating human and cyber-physical systems represents a new kind of production relationship. Only by establishing the production relationship between human intelligence and machine intelligence driven together by data and knowledge can it be possible to fully explore and release the enormously productive and sustainable forces potential embedded in energy and transportation integration.

The traditional artificial intelligence that relies purely on big data lacks explainability and is unsustainable due to increased energy consumption. Therefore the key to **4N4F** is to break through the sustainable development problems encountered in the process of smart energy and transportation. One is the explainability and autonomy of artificial intelligence, and the other is the sustainable energy demand of artificial intelligence.

The **4N4F** can solve the explainability problem of artificial intelligence by fusing perceptual and cognitive intelligence. In addition, it can realize the self-learning problem of artificial intelligence by continuously updating data intelligence to knowledge intelligence. Besides, the **4N4F** can reduce the data requirements of artificial intelligence by integrating data intelligence and knowledge intelligence. Furthermore, the **4N4F** decreases the energy consumption requirements of data processing by the small data/big knowledge structure based on edge computing and the decentralized clean energy/low-carbon transportation. These advantages of 4N4F help realize the future sustainable development of artificial intelligence (Fig. 3−21).

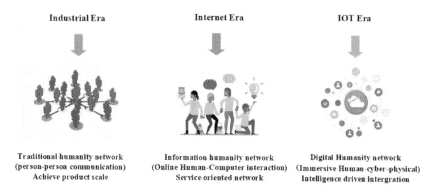

Figure 3−21 The influence of collaboration in different eras on the evolution of the humanity network.

In the past 200 years, during the industrial era, enterprises acquired a competitive advantage from the product scale and the direct communication between people. Moreover, in the past two decades during the Internet era, the competitive advantage became the network effect based on online human−computer interaction. In the future, the AI era, the competitive advantage comes from the ecological fusion effect based on the immersive human-cyber-physical integration.

As mentioned before, the core of 4N4F is integrating the humanity network under the human-cyber-physical ecosystem, which embodies the top-level design method of the digital economy. 5G technology provides technical support to promote the integration of the real and virtual world and the deep integration between networks. The humanity network supported by 5G can be embodied in smart industry, energy, transportation, and cities.

3.4.2.1 The embodiment of the humanity network in the smart energy operating system

First, the smart energy operating system is equivalent to the endogenous basic framework and rules of the sustainable development of the ecosystem, a platform for integrating human and machine intelligence, a digital mapping and carrier of integrating human-cyber-physical, and an engine to realize smart energy. Second, due to the self-similar structure and characteristics of the smart energy operating system, human−machine intelligence integration can be realized on the edge or the cloud side. Third, the realization basis of the intelligent energy system is the deep integration of 5G, cloud computing, artificial, and knowledge intelligence. Finally, the 4N4F ecological integration innovation with the trinity of the human-information-physical system is the key to the smart energy industry in the future.

3.4.2.2 The embodiment of the humanity network in the smart transportation

Smart transportation is defined as realizing the smooth interconnection of different transportation methods through the IoT to alleviate urban traffic, provide the most personalized intelligent transportation services, and satisfy participants' needs in the industry chain. To achieve such functions, we should value travel as a service, where people must be the core, and door-to-door travel intelligent management is essential. Door-to-door travel can connect sea, land, and air and meet sustainable development goals. The realization of such Travel as a Service requires the integration of four travel elements, namely the need (traffic users), carriers (manufacturers), routes (government regulations), and conditions (operating environment).

3.4.2.3 The embodiment of the humanity network in the smart industry

A large amount of knowledge and models accumulated in the industrial field for many years can solve qualitative problems well. However, these models cannot accurately match the fluctuating working conditions in many scenarios, and the industrial process is still a black box. In addition, tacit knowledge possessed by factory masters and craftsmen needs to be inherited and reproduced. In order to solve the above problems, we should use industrial intelligence for the intelligent empowerment of the industry. Through the intelligent cognitive engine driven by the knowledge graph, the smart prediction engine via the AI model, and the decision optimization engine through operations research and planning, the cyber-physical system-driven digital twin can evolve into a cognitive twin driven by the human-cyber-physical system. These engines will allow many technologies that were difficult to apply in the past to be embedded into industrial scenarios, promoting the empowerment of industrial intelligence.

3.4.2.4 The embodiment of the humanity network in the smart city

Urban intelligent computing involves specific research issues, challenges, architecture, methodology, and application scenarios in complex systems and emphasizes the use of big data and spatial-temporal semantic graphs to solve various specific problems cities face. The smart city is related to the future quality of life and sustainable development of human beings. It is also the starting point and leading strategic topic for the future national development of artificial intelligence. The smart city's fundamental framework of intelligent computing includes multiple links of perception semantics, data management knowledge, data analysis via graph, and service integration.

Among them, perception semantics is essential. We can update it in two ways. The first is to continuously track the status patterns of related subjects by perception platforms that connect people, things, processes, and applications and form the trajectory data of related subjects. The second is to collect information posted on social media.

1. If these copious amounts of unstructured data can be semantization through the city semantic information model, the large-scale trajectory and social media data generated by related subjects can be efficiently organized and managed for subsequent real-time analysis and mining.

2. When an abnormality occurs in a city, we can determine the spatial range and time interval more accurately based on the semantic data of these trajectories. The reason is that when an abnormality occurs, the subject's status attributes and the related parties' choices and feedback will change.

3. We can use social media or other business behaviors partly associated with these places and periods to analyze the causes of abnormalities.

4. Finally, this information will be delivered to the city management department and related parties authorized by the subject in time to quickly deal with abnormalities to avoid further state imbalance and provide a basis for future system decision-making.

3.4.3 The 4N4F drives the transformation from Industry 4.0 to Industry 5.0

The 4N4F emphasizes the integration of human-machine-environment to reconstruct the economy, society, culture, and environment that humans rely on for their survival. It represents the developmental trend from the fourth industrial revolution centered on AI to the fifth industrial revolution centered on man-machine-environment integration.

Explainable AI is the key to the fourth industrial revolution and digital economy. The digital humanity network in Industry 4.0 contains domain knowledge, industry norms, and market rules/business models, which require industry design methods and standards, mature operation, management experience, and data from stakeholders to converge knowledge and data. By introducing a digital humanity network, the 4N4F can produce a technical framework based on the five elements of data, algorithms, computing power, platform, and ecology. Such a framework will facilitate perceptual and cognitive intelligence interaction, ensuring the safety, security, and explainability of AI in energy and transportation. It will also overcome the barrier to the seamless integration of the mechanism model and machine learning while advocating the key concept of human-cyber-physical integration.

In a people-oriented smart society driven by the digital economy, the fourth industrial revolution fundamentally changed the interactions among science, engineering, and humanity. It also promoted further fusion of the humanity network with the infrastructure of three major revolutions in the physical world in three aspects: The energy revolution has changed the form of the physical world and made the sustainability of human life possible. The information revolution has changed the order of the physical world and shortened the temporal gap between people. The transportation revolution has changed the connection inside the physical world and reduced the spatial distance between people. With the advent of IoT, new production relationships have emerged between people and things,

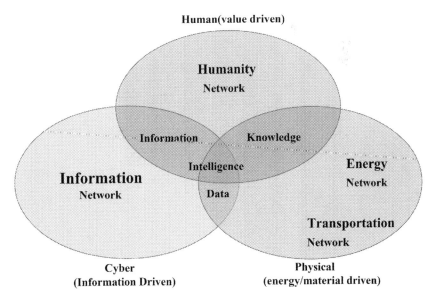

Figure 3–22 4N4F in the digital economy.

paving the way for the establishment of digital industry humanity in and out of the enterprise. Ecological innovation model based on user behavior-driven X as a Service emerged and accelerated the profound fusion of information, energy, transportation, and humanity network. The value of service effect from information, energy, material, and value flow is added to achieve the maximum benefits of the overall system.

The 4N4F in the digital economy is shown below in Fig. 3–22.

References

[1] L. Shidong, On the sixth Information Revolution, China N. Commun. 16 (14) (2014) 3–6.

[2] G. Chengshi, Digital transformation: from Information Internet to Metaverse, Zhang Jiang Sci. Technol. Rev. 3 (2022) 14–17.

[3] J. Chunlei, The future of the universe and wisdom city, J. Intell. Build. Intell. City 6 (2022) 153–155. Available from: https://doi.org/10.13655/j.carolcarrollnkiibci.2022.06.046.

[4] H. Xinju, D. Rui, W. Huaqing, Development and governance of the Chinese Metaverse: a comparison of the history of internet development in China [J/OL], Ind. Econ. Rev. (2022) 1–14. Available from: https://doi.org/10.19313/j.carolcarrollnkicn10-1223/f20220719.002.

[5] Y. Guoming, Z. Linyi, Future propagation from the perspective of the Metaverse: embedding and shaping of algorithms, Mod. Publ. 2 (2022) 12–18.

[6] H. Xu, Z. Li, Z. Li, et al. Metaverse Native Communication: A Blockchain and Spectrum Prospective, 2022.

[7] H. Xu, 5G opens the door to the metaverse, Zhang Jiang Sci. Technol. Rev. 2 (2022) 36–39.

[8] D. Qionghai, Meta-universe and Artificial Intelligence. "To Seize the Opportunity of the Development of the New Track, Create the Meta-universe to promote the New Bureau" Meta-Universe Summit Forum, 2022.

[9] D. Archer, The Long Thaw: How Humans Are Changing the Next 100,000 Years of Earth's Climate, Princeton Science Library, Princeton, N.J., 2016.

[10] H. Harvey, Designing Climate Solutions: A Policy Guide for Low-Carbon Energy, Island Press, Washington, D.C, 2018.

[11] D. Archer, The Long Thaw: How Humans Are Changing the Next 100,000 Years of Earth's Climate, Princeton Science Library, Princeton, N.J., 2016.

[12] A. Sayigh, Sustainable Energy Development and Innovation, Springer Press, 2022.

[13] K. Schwab, The Fourth Industrial Revolution, Portfolio Penguin, 2017.

[14] C.C. Chan, F.C. Chan, D. Tu, Energy and information correlation: towards sustainable energy, J. Int. Counc. Electr. Eng. 5 (1) (2015) 29–33.

[15] G.Y. Zhou, C.C. Chan, D. Zhang, et al., Smart energy evolution road-map based on the correlation between energy and information, Energy Procedia 158 (2019) 3082–3087.

[16] C.C. Chan, White Paper on 4 Networks and 4 Flows for Industry Integration Development, 2020.

[17] Y. Yang, R. Zhang, Study on Effect of Electric Vehicle Industry to Energy and Environment. International Conference on Advances in Energy and Environmental Science[C], 2013.

[18] J. Wu, X. Wang, Y. Dang, L. Zhihan, Digital twins and artificial intelligence in transportation infrastructure: classification, application, and future research directions, Comput. Electr. Eng. 101 (2022) 107983.

[19] L.-A. Carlos, R. Francisco, V. Francisco, Impact of digital transformation on the automotive industry, Technol. Forecast. Soc. Change 162 (2021) 120343.

[20] N. Saxena, S. Vibhandik, Tesla's competitive strategies and emerging markets challenges, IUP J. Brand. Manag. 18 (3) (2021) 57−72.

[21] E. Shokouhmand, A. Ghasemi, Stochastic optimal scheduling of electric vehicles charge/discharge modes of operation with the aim of microgrid flexibility and efficiency enhancement, Sustain. Energy Grids Netw. 32 (2022) 100929.

[22] B.K. Sovacool, J. Kester, L. Noel, G. Zarazua de Rubens, Actors, business models, and innovation activity systems for vehicle-to-grid (V2G) technology: a comprehensive review, Renew. Sustain. Energy Rev. 131 (2020) 109963.

[23] T.d.V. Saraiva, C.A.V. Campos, R.dos R. Fontes, C.E. Rothenberg, S. Sorour, S. Valaee, An application-driven framework for intelligent transportation systems using 5G network slicing, IEEE Trans. Intell. Transp. Syst. 22 (8) (2021) 5247−5260.

[24] S.M. Christina, P. Adrian, The role of 5G technology in sustainable development of smart cities, Ann. Dunarea de. Jos Univ. Galati. Fascicle I. Econ. Appl. Inform. 25 (2) (2019).

[25] C. Yan, S. Sheng, Research On Behavior Element Decision Of Cloud Edge Collaboration Capability Based On 5g Automatic Driving Scene. 2021 IEEE 5th Information Technology, Networking, Electronic and Automation Control Conference (ITNEC)[C], 2021.

[26] R. Preeti, S. Rohit, Intelligent transportation system for internet of vehicles based vehicular networks for smart cities, Comput. Electr. Eng. (2023) 105.

[27] Basic Concepts of Industry 1.0, 2.0, 3.0, 4.0. Standardization and Quality of Machinery Industry (12) (2015) 8.

[28] L. Wanyun, Y. Jun, Industrial 4.0 Concept Technology and Evolution Case [M], Tsinghua University Press, 2019.

[29] C. Cunbo, L. Jianhua, The connotation, architecture and enabling technology of industry 5.0, J. Mech. Eng. (2021) 1−13.

[30] N. Ma, X. Yao, F. Chen, H. Yu, Human-based intelligent manufacturing for Industry 5.0, J. Mech. Eng. (2022) 1−15.

CHAPTER FOUR

Strategy of integrated humanity for specific applications

Chaoqiang Jiang[1], C.C. Chan[2], Tianlu Ma[1], Jingchun Xiang[1], Xiaosheng Wang[1], Chen Chen[1], Hongbin Sun[3,4] and George You Zhou[5]

[1]Department of Electrical Engineering, City University of Hong Kong, Hong Kong
[2]The University of Hong Kong, Hong Kong SAR, P.R. China
[3]Department of Electrical Engineering, Tsinghua University, Beijing, P.R. China
[4]College of Electrical and Power Engineering, Taiyuan University of Technology, Taiyuan, P.R. China
[5]National Institute of Clean and Low–Carbon Energy, Beijing, P.R. China

Abstract

The core of the Four Networks Integration is the integration of the humanistic network under the "human-cyber-physical" ecosystem. This embodies the top-level design thinking for the development of the digital economy under the background of the new infrastructure, that is, the method of establishing the integrated development of human-cyber-physical systems. This is reflected in the applications of smart energy, smart industry, smart transportation, and smart cities. Specifically, in this chapter, the smart energy system including energy flow and information will be discussed; the smart transportation system relying on communications and the exchange of data between different platforms will be investigated; the collaborative robots for industrial Internet-of-Things in the smart industry system will be explored; and the ecosystem operating system for smart city with intelligent computing will be investigated. Consequently, it is shown that the integration of the humanistic network has become the key to accelerate different intelligent applications.

Keywords: Energy Internet; smart energy system; electrification; smart transportation; collaborative robots; industrial Internet; ecosystem; low carbon city; smart city

4.1 Energy Internet initiative

4.1.1 Energy Internet summary

Energy Internet (EI) is based on the existing energy networks, such as the power grid and gas networks, using advanced power electronics, information, and intelligent management technologies to realize the coordinated operation of renewable energy systems, distributed energy collection devices, distributed energy storage devices, and electric loads. With this huge combination and suitable regulation, EI acts as a platform to optimize the energy service delivery to end consumers [1,2].

As can be seen from Fig. 4−1, the EI is composed of four complex network systems, namely the power and transportation systems, together with the natural gas and information networks. These are closely coupled. First of all, the power system, as the hub of the mutual transformation of various energy sources, is the core of the EI. Secondly, electric power system and transportation system interact with electric vehicles through charging facilities. The layout of charging facilities and the driving and charging behaviors of car owners will affect the traffic network flow. On the contrary, traffic network flow will also affect the driving and charging behavior of car owners, and then affect the operation of the power system. Power-Gas (P2G) technology converts the excess output from

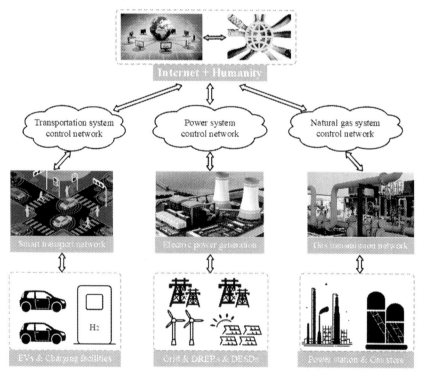

Figure 4−1 Main composition of Energy Internet.

renewable energy units into methane (the main component of natural gas), which is then injected into the natural gas network for transport and utilization. Further, the EI could also integrate other secondary energy networks such as heating networks. Thermal energy is an important by-product of distributed gas generation. The power network and the heating network can be integrated and coordinated with each other within the CHP (combined heat and power system) system, and the overall energy efficiency of the system can be greatly improved by using the waste heat discharged by the gas unit.

EI has the tremendous advantage of improving the comprehensive utilization rate of multiple energy sources such as electricity, weather heat, and solar energy. It will resolve some shortcomings of the traditional power and smart grids, such as inadequate participation of consumers, weak market mechanisms, inadequate utilization of energy forms, a dependency on existing structures, and security weaknesses.

The initial prototype of EI has been proposed for several decades, but it was first conceptualized theoretically in the meeting of FREEDM (The Future Renewable Electric Energy Delivery and Management Systems) in 2008 [3]. It was proposed that the advanced electronic technology and information technology should be merged into the existing power system to form the EI. It has five specific capabilities: (1) A new grid delivery interface that allows plug and play of any DRERs (distributed renewable energy resources) or DESDs (distributed energy storage devices), anywhere and anytime. (2) A communication infrastructure backbone that allows the real-time management of loads, DRERs, and DESDs through DGI (distributed grid intelligence). (3) The capability of being totally isolated from the main grid, if necessary, with automatic continuation of operation based on 100% renewable energy. (4) Perfect power quality and guaranteed system stability software. (5) Improved overall system efficiency, operating the alternating current system with unity power factor. In 2011 it was proposed in the book, the Third Industrial Revolution [4], that combining energy production, manufacture, storage, and transportation with the Internet is the most effective solution to the energy crisis. Within the blueprint of the EI, clean and renewable energy could be stored and shared like information on the Internet. Besides, every building and family could connect to this EI and can be self-sufficient while selling surplus energy, acting as a consumer and a producer. In 2015 a Chinese organization, the "Global Energy Interconnection Development and Cooperation Organization" (GEIDCO), founded the first dedicated organization to promote and encourage the sustainable development of a global EI [5]. The EI has great significance in the development of smart buildings and smart cities because it provides fundamental facilities for their operation [6].

4.1.2 Energy flow and information flow for smart automobile charging

Nowadays, both charging methods, namely slow and fast charging, can be operated normally regardless of the status of the power grid. However, these new mobile electric loads, with the noticeable increasing penetration of PEVs (plug-in EVs), will inevitably initiate great surges of demand in peak hours, causing negative effects toward the stability and security of the power grid. Fortunately, by carrying information into power, smart charging has been confirmed as one of the best solutions to the above-mentioned challenges. Upon the employment of smart charging, the real-time status of power grid and the demand of the PEV owners are monitored and analyzed. Consequently, the charging power can be adjusted with an online algorithm under the smart charging operation so that the conventional peak load problems can be suppressed. After that, the stability and efficiency as well as the average operating cost of the grid can be improved. According to the survey, with the adoption of the smart charging algorithm, EV smart charging can effectively reduce charging costs by 30%, grid operating costs by 10%, and renewable energy reduction by 40% [7].

The blueprint of smart charging within the smart grids is depicted in Fig. 4−2 [8]. The key to the smart charging is to communicate the information between all the parts, such as the PEVs, EV supply equipment (EVSE), regional power grid, and the control center, with the group. In the current smart charging algorithm, the EV batteries can return the excessive electric power back to the power grid, this is an essential feature of the vehicle-to-grid (V2G) technology. The realization of V2G ensures the reasonable regulation of energy flow and information flow between the smart grids and EVs. In addition, optimal charging references may need to be implemented with the EVSEs. The cooperation, collaboration, and coordination among several key players/stakeholders, namely, the automakers, electricity among several owners, and charging service providers, are the key to success toward the market penetration for PEVs.

The integration of the flows between energy and information is the key issue of EV smart charging [9]. The energy and information flow

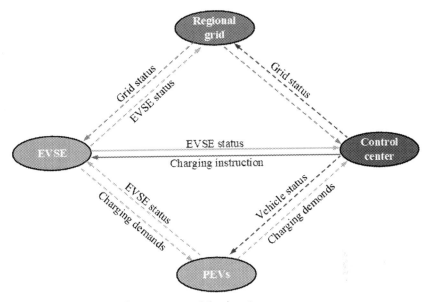

Figure 4–2 Blueprint of smart automobile charging.

among renewable energy generation, smart grid, advanced metering infrastructure, smart home, distributed network, microgrid, and EVs should be seamlessly correlated, as shown in Fig. 4–3. The flow of energy becomes bidirectional, showing the rich scenarios of V2X such as V2G, V2H, V2V, and V2B. At the same time, it can be considered to be integrated with communication base stations and data centers, and electricity can be sent along with the network, expanding the geographic scope and application scenarios of microgrids, and iteratively upgrading to serve more industries and individuals. In particular, the shared information can be spread to everyone effectively to improve its accessibility to potential readers. Merely producing more energy with more derived information is not enough, unless there is a change in the social perception of energy usage. For the applications of generation of energy, transmission, and distribution, the employment of control and energy utilization by information sharing has been proposed in the past few decades. With the advancement of information technologies, better control algorithms as well as more effective operations can be accomplished. The voltages and currents can be regarded as the derived information for the electrical systems, while the synchronized vector approach can be utilized to result in quicker and more accurate real-time responses.

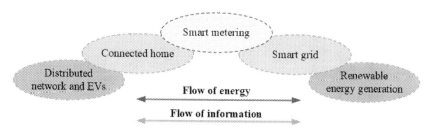

Figure 4–3 Integration of flows between energy and information in smart charging.

Upon any fault or malfunction within the electrical system, the change among the energy or information can be shown through the voltage drop or an increased current magnitude. The fault can then be detected by the protection system and the circuit breakers are activated to remedy the fault in a short period of time as requested. The spontaneous information can be shown by the change in energy so that the current status of the system can be achieved by monitoring the change of the information entropy. The information entropy in the functional system can be utilized to reach the optimization status in a healthy system.

Various energy information, such as transportation, working environment, preparation of food, and civic services, have been employed to provide energy consumption information for a metropolitan city. To be specific, big data can be utilized to serve as an important role in managing the whole city in a better way. For instance, optimization of the daily trips can achieve minimized energy consumption with the help of a global positioning system (GPS) on any vehicle. The improved platform and services can be achieved through the analysis of the behavior of citizens. Various aspects, such as sustainable energy employment, social harmony, and integration, can be accomplished with the rapid development in big data analysis and utilization.

4.1.3 Smart energy system

A smart energy system is defined as an approach in which smart electricity, thermal, and gas grids are combined with storage technologies and coordinated to identify synergies between them in order to achieve an optimal solution for each individual sector as well as for the overall energy system [10].

Compared to the smart grid, the smart energy system has taken an integrated holistic focus on the inclusion of more sectors, including electricity, heating, cooling, industry, buildings, and transportation. Besides, it also allows for the identification of more achievable and affordable solutions to the transformation into future renewable and sustainable energy solutions. The smart energy system driven by multidomain knowledge integration is shown in Fig. 4—4.

Through the establishment of a cross-disciplinary model and design method at energy ecosystem level, electric, thermal, and chemical energy are equivalently treated as energy products and services during joint planning stages, while renewable energy and fossil energy are coupled through hydrogen energy. A new industrial integration of multitype energy for the green chemical industry could be considered, and a large-scale hydrogen-based CHPF (combined heat, power, and fuel) poly-generation path implied from multifield knowledge fusion could be established. Among them, the system integration research based on high-temperature coelectrolysis technology has the potential to further promote the theoretical development of the energy ecosystem and has provided a technical basis for knowledge accumulation and optimization in engineering practices such as the implementation of IGFC (integrated gasification fuel cell) technology under a clean coal application strategy.

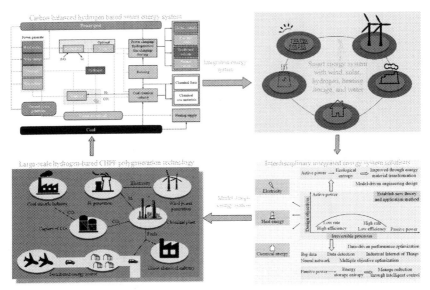

Figure 4—4 Smart energy system driven by multidomain knowledge integration.

By building a microgrid, the traditional energy model can be transformed into a decentralized and on-site new energy generation model. At the same time, local power generation as well as efficient use of green renewable energy put forward higher standards for stability and cost reduction on power transmission and distribution and bring many new challenges to the operation of the transmission grid and distribution network. Based on the scientific research and operational scheduling experience of the power system, an independent distribution system operation center (DSO) responsible for regional energy scheduling can be built as a key mechanism to coordinate the operation of the power system distribution network and adapt to the instantaneous change from various factors in the microgrid, and in turn support the carbon certificate and energy trading based on blockchain technology. The DSO responsible for regional energy dispatch includes microgrid analysis, calculation, coordination, low-voltage user power monitoring, commercial insurance services, and other applications, which play a role in the consumption, distribution, and storage of distributed energy so that effective management and control can be carried out to obtain the maximum value of the microgrid.

For example, a typical microgrid application near the user side is the integration of solar, storage, and charging. The microgrid plays the role of energy producer and consumer at the same time. The internal microgrid realizes energy flow dispatching through the direct current and digitalization of the low-voltage application. The unified intelligent management of the distribution network and the charging network realizes the integrated intelligent coordination of "solar storage and charging," so that the charging and discharging of electric vehicles are more efficient and carbon emissions are lower. The outside of the microgrid interacts with other microgrids or upper energy grids to transmit energy supply and demand information and controls microgrid energy flow in temporal and spatial dimension. Based on AI's accurate estimation of market prices and trends, as well as continuous learning of the demand of load, it can make accurate decisions on future energy consumption and actively participate in the market of frequency regulation and peak shaving of the grid. The microgrid has also become a link connecting smart transportation and smart energy, intelligently integrating mobile energy storage on tens of millions of electric vehicles into the "Four Networks" supersystem. The distributed network planning of energy ecosystem is illustrated in Fig. 4—5.

By establishing a self-similar energy network structure that integrates with edge computing, the nanogrid research platform and the microgrid

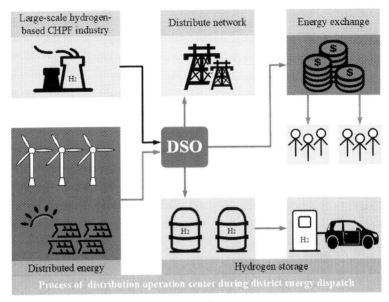

Figure 4–5 Distribution network planning of energy system.

demonstration project, the localized operation of energy services and the networking of energy systems can be designed based on the concept of the energy operating system (NICE_EOS) and the Nicer-Net. Nicer-Net is integrated and implemented through the architecture of virtual private cloud (NICE_CLOUD), edge computing (NICE_FOG), and device side gateway (NICE_MIST). Fig. 4–6 shows the Distributed Operation Design of Energy System.

Finally, through NICE_EOS and Nicer-Net, a knowledge base platform for the energy system can be established, which can be directly integrated with self-similar regional energy system operation optimization. Through interaction with system operation data, based on the energy ecological knowledge graph, data-information-knowledge-intelligence can be established for the evolution process of the carbon neutral energy system as shown in Fig. 4–7.

4.1.4 Integrated energy system

An integrated energy system (IES) is defined as a system in which various energy flows are interconnected and coordinated to release potential flexibility for more efficient and secure operation [11]. It is an important method to break down the barriers between different energy sectors in the EI. In the EI, electricity networks, heating networks, natural gas

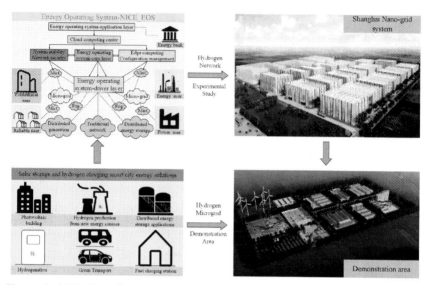

Figure 4-6 Distributed operation design of energy system.

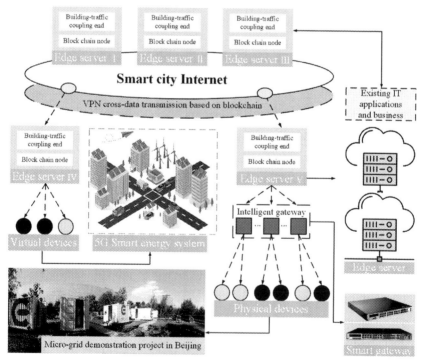

Figure 4-7 Implementation of distributed of carbon neutral energy system.

networks, and other energy networks are operated and managed coordinately to form so-called IESs. The interconnected operation and coordinated management of various energy sectors is shown in Fig. 4—8.

To coordinate different energy sectors and address the complication brought by the deep coupling between multiple energy networks, an integrated energy management system (IEMS), which is essential for controlling various energy flows in an IES, has been created to guarantee the safe and efficient functioning of IESs. The IEMS is viewed as the brain of the IES, and its main components are shown in Fig. 4—9. In IEMS, supervisory control and data acquisition controls the asynchronized real-time measurement data from various energy networks. The measurement data is then processed by dynamic state estimation to perceive IES states. To identify potential security threats, energy flow analysis and security assessment and control are

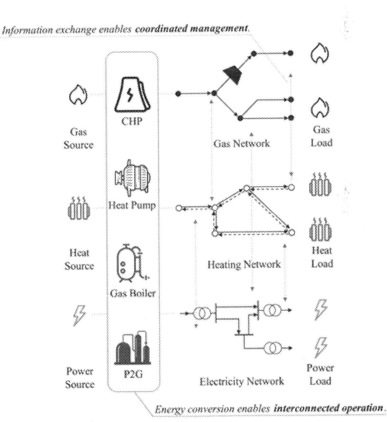

Figure 4—8 Interconnection and coordination of energy sectors in integrated energy management system [12].

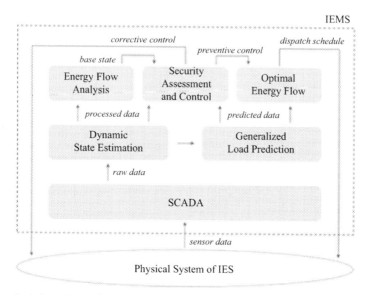

Figure 4—9 Function architectures of integrated energy management systems [12].

carried out by combining the estimated data with the future data from load prediction, which takes the correlation of multienergy loads into account. Furthermore, controllable resources in IESs are expected to be coordinated for more effective operation employing optimal energy flow, which takes advantage of the benefit of multienergy complementation.

The various time scales in different energy sectors pose a great challenge to the operation of the IEMS. The transient processes in electricity networks usually have a time scale of milliseconds, while the transient processes in other energy networks of IESs, such as natural gas networks and heating networks, can last minutes to hours so that they cannot be omitted in the normal operation of IESs. So, the IEMS needs to establish its modules based on a dynamic IES model, which is described by both partial differential equations and algebraic equations. Inspired by the electric circuit modeling of electricity networks, an energy circuit method (ECM) as shown in Fig. 4—10 that models natural gas networks and heating networks in the frequency domain is built as a key mechanism to solve the challenge [12]. The ECM is a circuit-analog model for IESs that fully draws on the deduction of electric circuit modeling from the time domain to the frequency domain and from distributed parameters to lump parameters. This model accurately reflects the dynamic processes in IESs and is tractable enough for calculation in the modules of IEMSs.

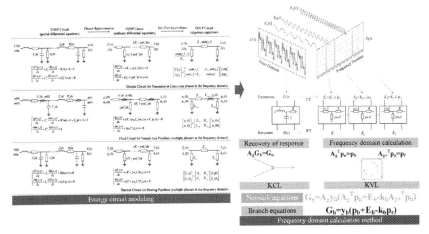

Figure 4—10 Energy circuit method description [12].

4.2 Intelligence connected transportation

The transportation industry is important for a country, which specializes in transporting goods and passengers, including land-based, water-based, and air-based transportation carriers. Nowadays, the advances in artificial intelligence (AI), big data, smart sensors, and renewable energy benefit the development of transportation and traditional travel modes are profoundly changing. The human beings' needs for daily life with high efficiency and high quality also ask for big evolution in transportation. Meanwhile, intelligence-connected transportation can reduce greenhouse gas (GHG) emissions while it also benefits environmental protection and energy security. All these demands will promote the next generation's smart transportation system.

4.2.1 Electrification, enabling green travel

The development of technology and science services for human beings improves the efficiency and comfort of daily life. The three industrial revolutions have greatly liberated labor productivity. Meanwhile, it also lays the hidden danger for environmental issues. Climate change, which is led by GHG emissions, is a big challenge for human beings, and the transportation sector makes one of the major contributions to GHG emissions. Oil is still the main source of energy in our daily lives. However, clean energy will become dominant in several decades, according to the forecast about the energy market [13]. In recent years, renewable energy has attracted attention all over the world. Solar photovoltaics and wind power are both representations of sustainable energy. The proportion of sustainable energy in the grid is going up year by year. These kinds of energy are all stored in the form of electricity and therefore preparing for the electrification process.

During this present period, the electrification of transportation will play a significant role. Transportation electrification includes evolutions in land-based, water-based, and air-based tools. In future the green travel system will be established. Fig. 4—11 shows the trend of the proportion of GHG emissions from different modes of transport [14], shipping and aviation have been accounting for a large proportion of GHG emissions, This will continue for quite a long time to come, and the proportion of traditional land-based transportation will reduce after 2030, electrification will help to fill in this missing part. To address climate change, many countries have published their own plan for reducing carbon emissions. For example, in 2019 the United Kingdom brought forward a ban on sales of fossil fuel cars from 2040 to 2035, and in 2021 the Chinese government promised China would strive to achieve carbon peaking by 2030 and carbon neutrality by 2060.

In addition, there are also many initiatives to realize transportation electrification. Infrastructure construction is the basis of transportation electrification. On the one hand, the plants of solar power, wind power, and other renewable energy need to be arranged to make sure the share of sustainable energy gets stable growth. The ratio of renewable energy

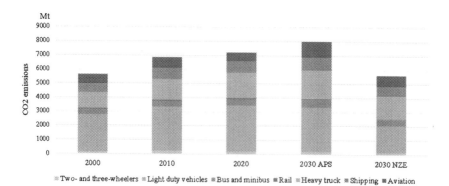

Figure 4—11 Global CO_2 emissions from transport by subsector, 2000—30 [14].

would determine GHG emissions, which is a sign of success in electrification. A free fuel-based power network is the ultimate target of transportation electrification. Then, charging piles and charging stations should be promoted and popularized, just like gas stations are located in various areas in daily life now. They will need to expand to suit the growth of the electric vehicle markets [15]. Fig. 4–12 shows the trends in charging infrastructure from 2015 to 2021, the number of both fast and slow publicly available chargers, have been increasing exponentially. Since the development of electric vehicles is still in its primary stage, this rapid

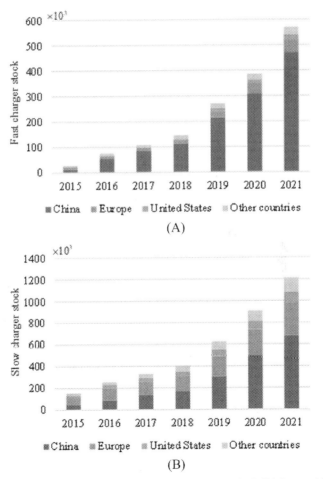

Figure 4–12 Trends in charging infrastructure, 2015–21 [15]. (A) Fast publicly available chargers, 2015–21. (B) Slow publicly available chargers, 2015–21.

growth in infrastructure construction will indeed be maintained for a long time. What is more, the energy management will optimize the electricity distribution, since the energy sharing between different electrical terminals can promote grid balance, and the traditional electrical equipment can also be used as a source of power, which can feedback energy to the grid or other devices. Vehicle-to-Home, Vehicle-to-Vehicle, and Vehicle-to-Grid have been proposed to balance energy distribution [16]. They will effectively improve grid stability and reduce energy waste.

Meanwhile, there are also multiple challenges during the process of transportation electrification [17]. The first is to investigate a suitable rechargeable energy storage system. Currently, batteries are popularized in various fields. The battery with long cycle life, high energy density, and high power density still needs study and development. Compared to traditional fuel cars, electric vehicles (EVs) are limited by power batteries. The energy density of batteries and petrol is 200 Wh/kg and 12 kWh/kg, respectively. Thus the driving range of EVs is much less than that of fuel vehicles. This problem can also be solved by dynamic wireless power transfer (DWPT) technologies. Namely, EVs would be charged while being driven on the road, such charging facilities will cost a lot to build and maintain. And the performance of batteries is affected by temperature, which limits the expansion of EVs in high-latitude areas. Another concerning issue is safety. The reports on EV explosions can often be seen in the media. Therefore fuel vehicles are much more robust than EVs based on current technologies, and the breakthrough in batteries will greatly facilitate the electrification process.

The second challenge is related to energy management and balance. With the increase in sustainable energy, grid stability will face many problems. Sustainable energy generation relies on the environment, and the weather will affect the production of solar and wind power. Therefore the demand and supply might mismatch; the production of renewable energy might be very low in bad weather conditions while the electric equipment still needs the power to work. Therefore energy storage technology is a great approach to tackle grid stability. In addition, creating smart grids is also an effective way to handle this problem; energy balance will be optimized in multiple aspects based on AI and big data technology.

The third is the challenge of new technology, such as self-driving cars and cyber security, giving electrification services to human beings but handing over work to machines. Self-driving cars will become a part of daily life, where the control depends on code, not humans. Once cyber security is threatened, vehicles will go out of control. In addition, the uniform form of electrification transportation needs standardization in the industry. There are many organizations publishing standards, for example, the society of automotive engineers, the international electrotechnical commission, and the standardization administration in China. The uniform standards for the whole industry are meaningful and significant for transportation electrification; they can promote development while reducing waste. Cooperation among countries is also necessary for uniform standards.

4.2.2 Intelligence connected transportation

Intelligence transportation relies on communication and the exchange of data between different platforms. Sensors are the eyes through which the machine looks at the world, and they can obtain, collect, and process information from the surroundings. The intelligence transportation system depends on accurate data from sensors. The sensors can be divided into two categories according to their position. The first kind of sensor is installed in vehicles, and senses the information around it, and the data collected from the environment will be used for vehicle control. The other category is the sensor on the infrastructure, installed on stationary facilities, and they sense the information around themselves. EVs or other devices will be sensed while they are accessible to the sensing range. Sensors are essential for information collection, and the collected information will be transferred to control systems for further analysis or actions. More sensors make transportation smarter. There are three stages for intelligence-connected transportation.

Stage 1: Dynamic perception

Dynamic perception is the feature of intelligence-connected transportation; it is not only reflected in a large number of sensors but also in strong data analysis through new technologies such as AI. Take the vehicle as an example, the sensors on one vehicle can be as many as 200, and they can be classified by functions, and the detailed classifications and descriptions are shown in Table 4—1. The in-vehicle sensors can help monitor the vehicle's state and environmental information around the vehicle. On the other hand, in-road sensors are also important. The traffic data in real-time can be collected by the sensors along the road. These data can be analyzed by big data and AI, the congestion can be solved,

Table 4–1 Classification of sensors in vehicles [18].

Sensor functions	Sensor classifications	Descriptions
Safety	Distance sensors	To detect obstacles, hazards, people, or even other vehicles, including long- and short-range distance sensors.
	Night vision sensors	To assist driver's perceptions by sensing road and roadside.
	Speed sensors	To measure vehicle speed.
	Angular rate/linear acceleration inertial sensors	To maintain a vehicle's stability.
	Passive safety-support sensors	To improve the safety level of passengers.
	Positioning/navigation systems	To provide position and time information, like GPS.
Diagnostics	Sensors for powertrain diagnostics	To monitor the state of charge of the battery.
	Sensors for chassis diagnostics	To monitor tire pressure.
	Sensors for body diagnostics	To diagnose the compartment's electronics and ambiance.
Convenience	In-cabin convenience sensors	Monitor air quality, humidity, etc.
	Driving convenience sensors	Assist driver to drive.
Environment monitoring	Sensors for environment monitoring	Monitor environment information.

and efficiency can be improved. The in-road sensors can also be divided into two types, intrusive and nonintrusive. The intrusive sensors, like some magnetic sensors, and inductive loop detectors are installed on the highway surfaces. For example, the DWPT system can charge EVs while they are driven on the road, and the intrusive sensors can feel the positions and state of charge of EVs. However, installation and maintenance costs are higher than nonintrusive sensors. Nonintrusive sensors, such as radar sensors, video cameras, and ultrasonic, are installed at various locations along the road, they might be affected by weather conditions, but they can be easy to replace and repair. The investment in transportation infrastructure is essential for the development of a country, and the accuracy and robustness of sensors are hoped to promote.

Stage 2: Active management

Traffic management will become more based on automated means to sense, transmit, and apply information in the future, from passive "you go to see" to actively encouraging people to pay attention to and use information so as to improve their own operation management service capability. These are what "Intelligence Connected Transportation" hopes to solve.

The huge data collected from the in-vehicle and in-road sensors can be processed and analyzed further. On the one hand, the analytics of such huge data can be used for planning next-generation transportation networks. On the other hand, such huge data can help to manage transportation and optimize paths, then improve working efficiency. The optimal solution by the big data can offer active traffic management and active command and control. The traffic jam is an annoying problem for drivers and passengers; this problem can be well addressed by mining the data of sensors. Through calculation and analysis, the issues can be anticipated in advance and solved by optimized algorithms.

Stage 3: Intelligent and connected

Intelligent transportation faces many new development opportunities and challenges. The overall development trend of the intelligent transportation system in the future is intelligent and connected. The development of vehicle-road cooperation, on the one hand, needs the progress of transportation. On the other hand, it puts forward the corresponding demand for infrastructure intelligence. The sensors are added and updated in the transportation and infrastructure, and the data will be summarized and analyzed together; namely, all the information is connected. The intelligence will be reflected when massive data are summed up together. Fig. 4—13 shows the case of intelligence-connected transportation based on EVs. The critical point is that every node's information is traceable and can be extracted and analyzed. The combination of multiple data will produce exponential analysis results.

The self-driving vehicles will play a key role in transportation in the future; they rely on intelligent computer systems to perform driving tasks independently, replacing human drivers in transport. At present, self-driving vehicles encounter difficulties mainly because the key technology has not yet broken through, manufacturing costs are too high, and there are no corresponding laws and regulations. There is no doubt that these problems will be solved in the near future, and eventually, self-driving vehicles will become a major player in transportation. Self-driving vehicles

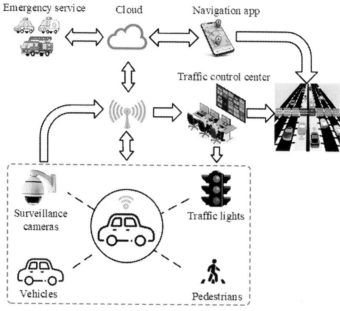

Figure 4–13 The case of intelligence-connected transportation based on electric vehicles.

sense the surrounding environment and road information through the onboard sensors, and they have freed humans from tedious and boring driving, which can not only significantly improve the efficiency of urban traffic but also effectively reduce or even avoid traffic accidents. They depend on accurate sensors and powerful computing capabilities, and AI and big data will help them integrate this intelligent transportation.

4.2.3 Smart transportation system

Smart transportation system services for travel of human beings, its core point is the "Mobility as a Service." And the intelligence of traffic depends on the interconnection of multiple vehicles through the Internet-of-Things (IoT). The application of smart transportation systems will mitigate traffic issues in the process of urbanization. Travel needs, travel carriers, travel routes, and travel conditions are key elements of the smart transportation system, and the integration of these four elements is the requirement to realize "Mobility as a Service." Meanwhile, the integration and connection between multiple transportation carriers will be various, with the requirements of participants with different needs.

"Mobility as a Service" is a people-oriented system; it needs intelligent management to achieve diverse and personalized services. It might be a complex combination of land-based, water-based, and air-based modes; all these schemes should follow the concept of sustainable development. Therefore this smart transportation has fundamentally changed the definition and value of traditional transportation carriers. Take the Internet of Vehicles as an example. Fig. 4—14 shows vehicle industry development through the digital economy to the passenger economy. It will finally be oriented to user needs, which is totally different from the traditional vehicle market. Therefore we must re-understand the real service needs behind passenger economy in the process of industrial chain integration.

In addition, the 5G brings new opportunities and challenges for the smart transportation system. The 5G Internet of Vehicles is not only the cooperation of vehicle and road but also the cooperation of people, vehicles, road, network, and the cloud. These five elements will be coupled with the smart transportation system. When multiple transportation carriers can be provided to solve the needs of passengers, the optimal path will be solved by AI and big data technology, and it still needs to integrate user needs, transportation, information, and energy networks. In the future, the vehicles will become computing nodes, which can share information with other nodes. During this process, the communication

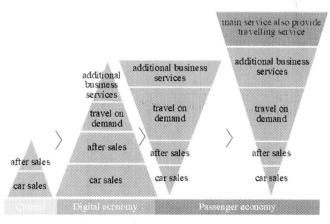

Figure 4—14 The core elements of a new generation of the smart transportation system [27].

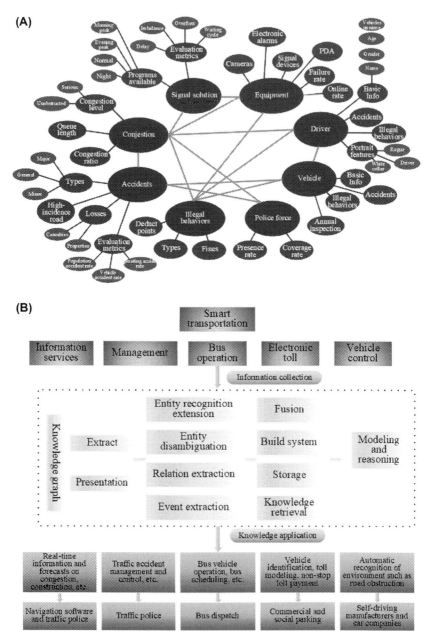

Figure 4–15 Knowledge of graphs drives the evolution of car networking operating systems to smart transportation [28]. (A) Urban transportation knowledge map. (B) Application framework of smart transportation knowledge graph.

between vehicles, people, roads, networks, and the cloud is significant and inevitable. And the smart IoT platform will be established, and it will collect and analyze information from multiple facilities. A powerful cloud computing platform also needs to be built to service the smart transportation system. To achieve "Mobility as a Service," safety is also an important point. The traffic flow and information flow need to be integrated.

Simple information interaction cannot utilize the advantages of new technologies, like big data, cloud computing, and AI. Therefore we need to create a new smart transportation system by integrating and sorting information from car networking. Fig. 4−15 shows the scheme that smart transportation needs to reclassify data for computational analysis. The distributed node information is processed in a unified manner to serve the information service, traffic management, bus operation, electronic toll, and vehicle control. The evolution of smart transportation needs to solve roadside problems and comprehensively solve the problems of interconnection between people, vehicles, roads, and the environment. The "Mobility as a Service" reflects that the core of smart transportation is to meet the needs of human beings, and the establishment of the smart transportation system needs multiindustry integration. The coordinated development of the "people-vehicle-road-network-cloud" will promote smart transportation to explore personal services, industry services, and public management services.

4.3 Collaborative robots for industrial Internet-of-Things

4.3.1 Industrial Internet-of-Things

The industrial Internet-of-Things (IIoT) can be seen as a subset of IoT which is applied in specific industrial applications. The definition of the IIoT is not fixed at present but is still gradually being improved by different scholars and institutions. The definition in Ref. [19] can provide more details.

"A system comprising networked smart objects, cyber-physical assets, associated generic information technologies, and optional cloud or edge computing platforms, which enable real-time, intelligent, and autonomous access, collection, analysis, communications, and exchange of process, product and/or service information, within the industrial environment, to optimize overall production value. This value may include improving product or service delivery, boosting productivity, reducing labor costs, reducing energy consumption, and reducing the build-to-order cycle."

The IIoT reference architecture provides guidance in the development and deployment of systems, which is a comprehensive and generic guideline, not every detail needs to be implemented in every application. System architects can use the reference architecture as a template to achieve a consistent architecture across different systems, meeting unique system requirements. This can also significantly enhance system interoperability among different industrial sectors.

The Industrial Internet Alliance released a document called Reference Architecture (IIRA) [20]. The core of IIRA is a set of system conceptualization tools called "viewpoints." The system requirements are divided into four viewpoints, which include business, usage, functional, and implementation views. The implementation viewpoint described the technologies and the system components that are acquired by the usage and implementation viewpoint. Except for IIRA, there are three main reference architectures, which are RAMI 4.0, IIRA, and OpenFog RA, respectively. An overview of the three reference architectures and some comparisons are presented in Ref. [21].

Fig. 4−16 presents the widely accepted three-tier IIoT system architecture [22]. The three-tier model is composed of edge, platform, and enterprise tiers. The edge tier, which usually consists of sensors, controllers, and actuators, collects data through the proximity network and sends data to the platform tier across the access network. As a middle tier, the platform tier provides nondomain services such as data query and

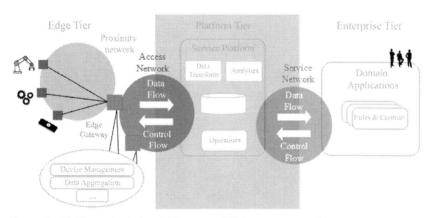

Figure 4−16 Three-tier industrial Internet-of-Things system architecture.

analytics. It analyzes the data received from the edge tier and transmits it to the enterprise tier through the service network. Similarly, the platform tier implements the control instructions of the enterprise tier and controls the operation of the edge tier. The enterprise tier provides the domain applications and interfaces to end users.

Although the system architectures provide the basic guide to the IIoT applications, the challenges caused by diversified industrial uses cannot be easily solved by only following the architectures. The key requirements of the IIoT and the emerging technologies used to address these challenges are discussed in Ref. [23]. Table 4−2 presents the relationship between emerging technologies and key requirements. Taking blockchain technology as an example, it addresses the security and privacy challenges in the IIoT, and interoperability, scalability, and reliability challenges can be further solved. Similarly, 5G technology can achieve low latency and solve the scalability and reliability challenges. With the development of these emerging technologies, each is expected to solve more challenges.

4.3.2 Collaborative robots in Internet-of-Things

As a typical actuator in the edge tier of IIoT architecture, robots play an important role in many applications, especially in the manufacturing industry. Industrial robots have been developing for more than half a century, and the technology is quite mature. The traditional industrial robot is usually an independent individual, carrying out specific manufacturing tasks according to preset instructions. Due to the lack of autonomous consciousness, industrial robots usually work in an enclosed space like a cage. Compared to traditional industrial robots, the most significant feature of

Table 4−2 Relationship between emerging technologies and the key requirements.

Emerging technologies	Key requirement
Edge/Fog computing	Interoperability, scalability, security, privacy, reliability, low latency
Software-defined networking	Interoperability, scalability, security, reliability, low latency
Blockchain	Interoperability, scalability, security, privacy, reliability
Machine learning	Privacy, security
5G technology	Scalability, reliability, low latency
Wireless Sensor networks	scalability, reliability, low latency

cooperative robots is that they can cooperate with people and have a sufficient safety performance [24,25].

The features of collaborative robots and noncollaborative robots are summarized in Fig. 4—17 [26]. Collaborative robots have decisional autonomy. For example, a collaborative robot equipped with an electronic safety skin can replan its path when it detects that something interferes with its original working path. For movements and workspaces, the collaborative robot can share workspaces and cooperate safely. In terms of communication, the collaborative robot supports real-time interactions. Besides, the collaborative robot is light-duty and easy to program, and deployment does not require specialized programming capabilities, as it provides more convenient drag-teaching and graphical programming functions.

Due to the characteristics of lightweight, easy deployment, and ability to work with people, cooperative robots are widely used in multiple industries. Fig. 4—18 presents the typical application of collaborative robots. Picking and placing, packaging, and palletizing are the most common application of collaborative robots. In most application scenarios, robots will move and place the workpiece from one place to another. Gluing, dispensing, and welding are also major applications of collaborative robots. Especially in the automotive manufacturing industry, collaborative robots can do welding on their own or with other workers. They can also help people assemble items and implement product quality inspection, improving efficiency and work quality. In some dangerous

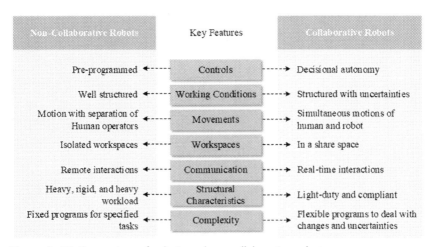

Figure 4—17 Comparison of cobots and noncollaborative robots.

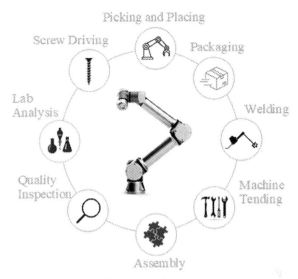

Figure 4−18 Applications of collaborative robots.

working conditions, such as handling harmful materials and analyzing dangerous reagents, collaborative robots can provide operators with a safe working environment. Similarly, humans can be liberated from boring and repetitive work with their help.

Safety is most important for robots to be able to work with humans. Collaborative robots interact with people, this evolution means existing security procedures and workspace boundaries need to be removed. Due to the high-speed movement and the strong or sharp workpiece carried by the robot, it may cause serious damage once a collision occurs. The open issues about the safety of industrial collaborative robotics are summarized in Fig. 4−19 [29].

The security issues of cooperative robots can be summarized into eight subcategories, which are not independent but sometimes overlap. For example, the research about robot skin can be classified as the design of intrinsically safe collaborative robotics or precollision approaches. The design of intrinsically safe robots is mainly about using some advanced control strategies and some external sensors or increasing some buffer devices to improve the safety of robots. The precollision approaches refer to how to stop or modify its trajectory to avoid collision when the robots sense humans or obstacles. In contrast, postcollision approaches study how to reduce the effects of a collision after it has occurred. The combination

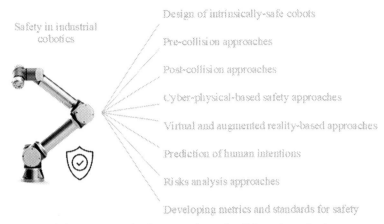

Figure 4–19 Subcategories of safety in industrial collaborative robotics.

and interaction of different sensors implement the security method based on cyber-physical systems. Virtual and augmented reality-based methods can simulate industrial cooperative robot systems with a high-level immersion, so that robots can detect potential collisions in advance and take timely action before any injury occurs. Robots also need to be able to predict human intentions, which will not only increase the safety of human-robot collaboration but also increase work efficiency and productivity. In addition, the research of risk analysis and the formulation of safety standards are also important issues.

Although collaborative robots with self-learning and self-decision ability will limit the human role, it does not mean that humans will be replaced by robots. With the development of the IIoT, the value of robots will be further developed. They will no longer be limited to repetitive and fixed tasks, and they will be able to combine their skills with human skills to complete more complex tasks. The relationship between robots and humans is cooperation, not competition. Furthermore, humans assume the role of leadership and supervision because there is no substitute for human creativity and flexibility. Humans will focus on information and data management, new technologies, and interaction with complex systems.

4.3.3 Smart industry system

Smart industry is the integration of the trinity of people, industrial information systems, and industrial physical systems. Intelligent factories are the typical application of the intelligent industrial system. Compared to traditional factories, intelligent factories can finish customized tasks using distributed networks [30]. The system architecture of the smart factory is shown in Fig. 4—20. The flexibility of robots is very important in smart factories. As the most low level actuators, robots can plug and play and can be reconfigured according to different requirements and processes. They are usually developed into components, which are the foundation of the system. At the middle level, there are many functionalities developed in the internal cloud, which are human-system interface, storage management, task planning, virtual manufacturing, and big data collection. Customers can make personalized customization according to their own needs, such as color, size, shape, and process. These customization requests are transmitted to the factory through the human-system interface. These orders are processed internally by the smart factory and are eventually completed by the actuators at the low level. The manufacturing execution system, sale management system, and design support system constitute the upper level.

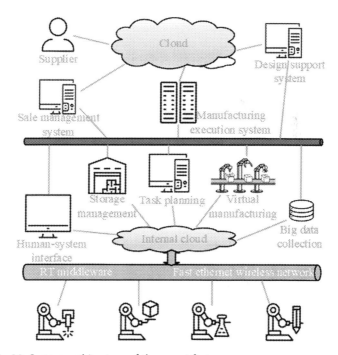

Figure 4—20 System architecture of the smart factory.

From the perspective of the evolution of the industrial value chain, the value of the smart industry is changing from a product-centric model in the past to a scenario-centric as-a-service ecological model. To achieve this goal, industrial Internet applications that integrate knowledge graphs have become an important innovative means. Among them, the establishment of a sustainable and updated knowledge base for subindustries through the integration of industrial humanistic networks will become the key to accelerating intelligent applications for smart industry. Industrial Internet architecture based on knowledge integration is shown in Fig. 4—21.

As shown in Table 4—3, the data center is also a typical application scenario. In the traditional mode, energy flow and information flow are independent, and there is a lot of room for improvement. After the integration of multiple flows, the information flow provides an optimization strategy for energy flow from a global perspective. Real-time "health examination" is carried out on the equipment in the loop to achieve a more reliable operation guarantee. During the operation of the data center, energy conservation and emission reduction are the most important evaluation indicators. Through information flow technologies such as artificial intelligence, the data center operation information is collected, analyzed, and dynamically modeled, and the best cooling strategy and power supply strategy are derived in real-time. At the same time, it combines the computing power requirements of the upper data flow to achieve the best overall energy efficiency of the data

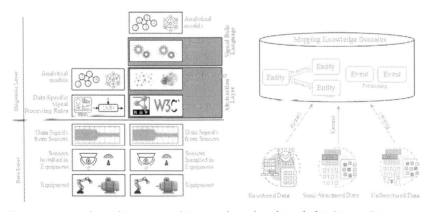

Figure 4—21 Industrial Internet architecture based on knowledge integration.

Table 4–3 Autonomous driving mode for data center.

Maintenance	Operation	Levels
Completely rely on human, manual recording	No optimization at all	L0
Electronic and digital operation and maintenance data, reducing manual inspection content	Basic monitoring and visualization of energy efficiency	L1
Recognize the status of static devices through AI technology, basically eliminating routine inspections	Rule-based PUE optimization	L2
Use AI technology to intelligently diagnose problems that occur during equipment operation, eliminate manual analysis	Automatic optimization of PUE energy efficiency based on AI, manual intervention	L3
Predict the health status of equipment through AI technology and preventive maintenance	Energy efficiency optimization based on reinforcement learning; business forecast, assist business decision	L4
Fully autonomous operation, self-repair of system failures	No intervention is required and the optimal operating state is automatically reached	L5

PUE, Power usage efficiency.

center. In the daily operation and maintenance process of the data center, AI can also be used to perform predictive maintenance on key equipment in combination with operating data and, ultimately, achieve the "automatic driving" of the data center.

The large amount of industry knowledge accumulated in the industrial field for many years can solve qualitative problems well. However, in many scenarios, these mechanism models cannot accurately match the fluctuation of working conditions, and the industrial process is still a "black box." In addition, the large amount of tacit knowledge possessed by factory masters and craftsmen needs to be inherited and reproduced. To solve the above problems, industrial intelligence has become a new engine for intelligent empowerment of the industry. Through the intelligent cognitive engine driven by the knowledge graph, the intelligent prediction engine via the AI model and the decision optimization engine through operations research and planning, the cyber-physical system driven digital twin can evolve into a cognitive twin driven by the human-cyber-physical system, which allows many technologies, that were difficult to be applied in the past, to be embedded into industrial scenarios and promotes the empowerment of industrial intelligence.

4.4 Ecosystem operating system for smart city

4.4.1 Ecosystem in smart city

Smart city is an open concept that aims to present more intelligent, interconnected, and efficient services to the citizens and organizations to face the challenge of sustainability, security, and urbanization. Smart city uses information and communications technology (ICT) to monitor, analyze, and govern all its critical infrastructures to achieve well-performance in terms of economy, health, mobility, and government administration as shown in Fig. 4—22. It is built on the combination of the endowments and activities of self-determined independent and conscious citizens [31,32]. Nowadays, metropolitan cities in both developed and developing countries (such as the Netherlands and India) have already adopted it for the purpose of improving the quality of citizens' lives and urban development [33]. The idea of smart cities was first proposed in the late 1960s in Los Angeles, and after 50 years of evolution, it came to the stage in European Union [34].

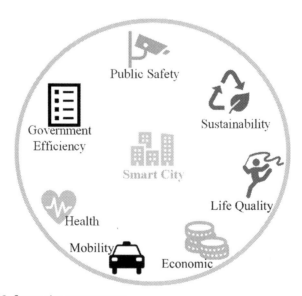

Figure 4—22 Smart city components.

Smart city utilizes information in its physical structure to optimize resources effectively and enhance the collaboration of entities and domains. The integration of different domains is a principal core in constructing smart cities. The domains could be classified into four types that are technological, governmental, and institutional as well as transitional components. There are various actors that are identified as universities, industries, and governments that exist in or interact with these components in an ecosystem, and a smart city consists of diverse ecosystems. Each ecosystem contains a variety of partners and serves to solve specific problems through the collaboration of organizations and individuals. The ecosystem is formed under the consideration of the interrelations between critical infrastructures, governmental administrations, and society.

As mentioned by Nam and Pardo [35], the value creators in the ecosystem are classified into four types, namely, governments, business organizations, communities as well as residents, which create and consume the outcome of the smart city as shown in Fig. 4−22. Generally, when talking about the smart city, numerous authors focus on smart lighting, smart green energy, and smart transportation, which are services provided by the governments, and the other three are omitted or with less discussion. Businesses and organizations bring service to the economy and finance and gain outcomes for their stakeholders. Uber and Google Maps are typical smart businesses that improve the quality of life and solve the problem of mobility as well as economic development. University campuses, large mansions, airports, and smart buildings require both special and localized needs as the communities, and smart services are customized exclusively for the stakeholders. Both residents and individual citizens also may bring out smart services for cities. Residents could provide live information through their sensors to monitor the traffic, air pollution as well as public safety to both governments and other residents.

The framework of the ecosystem in the smart city is shown in Fig. 4−23, this consists of multiple layers with a distinguished capacity. There is no priority in the six capacity layers, and each one plays a unique role to process the mission in the smart city. These layers are integrated through the common ICT to work in coordination with each other [36−38].

- **Innovation layer**

    It is essential for vale creators to update and make innovations to always update their services for their stakeholders. Smart cities

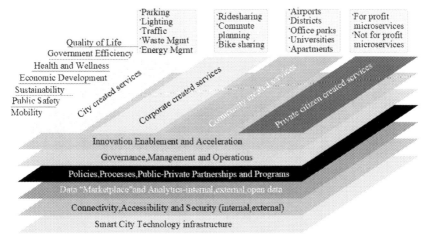

Figure 4–23 The framework of the smart city ecosystem.

strengthen the cooperation between government, enterprises, and campuses to carry out a variety of innovative programs, including workshops, innovation seminars, and skill training.

- **Governance and management layer**

  Smart cities promote conventional processes and services on a digital path. The management model monetizes the advanced business pattern with diverse support and plans for services and processes. The existing infrastructure will be updated and integrated with the fresh ecosystem that is renewed by the value creators and innovators. At the same time, a new set of criteria will be updated and released to refine smart city services.

- **Policy and financing layer**

  Smart cities developed with the accumulation of technology and capital. Building, operating, and maintaining smart cities require a model that engages standards, funding, and partners. Cities are required to evolve a new set of intelligent capabilities and awareness to keep on the stage in the "smart city."

- **Information and data layer**

  Smart city requires massive information and updated data as nutrients for sustainability. There are several ways to enhance the flow of information in the smart city, including sharing data initiatives and easing monetization policies. Ensuring information security and privacy is equally important as information collection and is a significant milestone for the smart city.

- **Connectivity, accessibility, and security layer**

  People, things, and networks are connecting seamlessly in smart cities, and there are still obstructions to managing, protecting, and verifying the connections and data transformation. Providing a seamless and trustworthy connectivity layer is a top priority for smart cities.

- **Innovation and technology infrastructure layer**

  The innovations and technology integrated into the infrastructure must be leading and overscale beyond the conventional municipal system, which could serve the new generation and future of value creators and stakeholders.

- **The utilization of ecosystem**

  As previously discussed, the smart city is an ecosystem of residents, governments, business groups, and institutions. There are other value creators involved that might work on their own or collaborate continuously. A sustainable and programmatic approach should be developed to take these stakeholders on a highway to engage with the smart city.

4.4.2 Low carbon smart city

The metropolitan cities face numerous challenges: overpopulation, power management, and public transportation stress. The increasing demand to live in cities can exacerbate sustainability challenges, climate issues, and energy allocation. Cities keep continuously attractive and competitive in global investment to generate jobs, income, and development funds, all of which lead to increased energy consumption and carbon emissions. New innovations and technologies are adopted to increase the efficiency of services and optimize energy distribution: smart transportation, smart and ultra-efficient buildings, and smart infrastructure, which will support low-carbon urban development.

Low-carbon cities and smart cities are two forms of frameworks for sustainable urban development, as shown in Fig. 4—24 [39]. The concept of low-carbon cities, which focus more on global warming and climate change, predates the introduction of smart cities. Smart cities benefit from the rapid development of information and communication techniques to enhance their capability to achieve a high-efficiency system. The low-carbon city concept is more about mitigation, while smart cities aim to adapt to the climate environment.

Figure 4—24 The framework for developing sustainable cities [39].

Smart cities can simultaneously transmit instant information and store big data, which enables governors to make systematic adjustments [40,41].

Low-carbon cities have clearer and more defined targets than smart cities, including sulfur, CO_2, and nitrogen oxide emission levels. However, the targets for smart cities are with fewer specifications, making the measurement and assessment of "smartness" more inaccurate. In addition, there are numerous research projects on low-carbon cities, while the research on low carbon in smart cities is relatively scant [42,43]. Several organizations and universities constructed various models and indicators to rank the performance of smart cities, such as the European Smart Cities Ranking (Fig. 4—25).

Smart cities go beyond low-carbon cities and transition to zero-carbon cities; the framework recommends considering compact city forms along with smart energy, especially smart grids, with ultra-efficient buildings, smart energy infrastructure, and clean electrification as three key considerations. Ultra-efficient buildings employ both low-carbon and high-performance materials for sustainability and set up smart microgrids with electrical systems, distributed energy sources, and smart management systems by analyzing demands to maximize efficiency. Smart energy infrastructure includes smart distribution grids and electric vehicle charging stations.

Clean electrification is electricity generated by green energy sources, such as wind turbines and solar panels. Numerous facilities such as energy storage units are set up for diverse applications including transportation, heating, lighting, and households to be operated with maximum clean electricity. The shape, size, density, and configuration of a network of

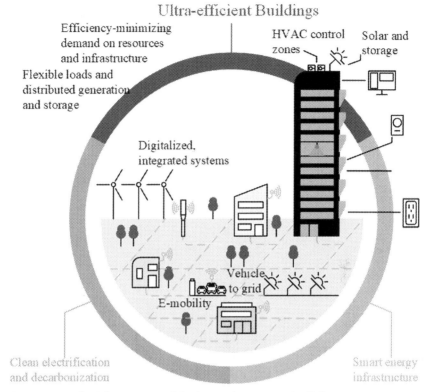

Figure 4—25 The infrastructure of the low-carbon smart city [44].

settlements are the physical characteristics used to describe the residential environment and are the basis for judging a compact city [44].

4.4.3 Smart city with intelligent computing

The construction of smart cities in the world has entered a period of rapid growth from the starting period. More and more systems, platforms, and data centers are being established, and all kinds of big data are quietly emerging in cities. The interactive mode of knowledge intelligence and data intelligence promoted by the integration of Four Networks also provides the possibility of urban intelligent computing in new smart cities. Urban intelligent computing extracts knowledge and intelligence through the integration, analysis, and mining of multisource heterogeneous data and combines industry knowledge to create a "human life-environment (energy/transport)-urban information" win-win situation [45].

The application system of urban intelligent computing is shown in Fig. 4–26. Urban intelligent computing involves specific research issues, challenges, architecture, methodology, and application scenarios of complex systems and emphasizes the use of urban big data and spatial-temporal semantic graphs to solve various specific problems faced by cities. It is related to the future quality of life and sustainable development of

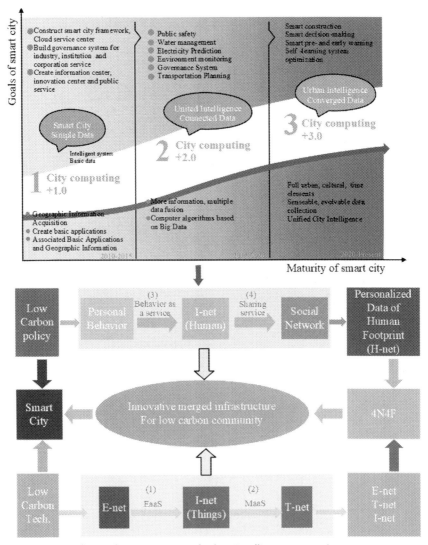

Figure 4–26 The application system of urban intelligent computing.

human beings, and it is also the starting point and strategic leading topic of the national future development of artificial intelligence. The basic framework of urban intelligent computing includes multiple links of urban perception semantics, urban data management knowledge, urban data analysis (via graph), and urban service integration.

1. At the level of urban perception, the system can continuously track the status patterns of related subjects by connecting people, things, processes, and applications through perception platforms that form the trajectory data of related subjects and can also collect information posted on social media for updates.

    a. If these large amounts of unstructured data can be processed through the city semantic information model, the large-scale trajectory and social media data generated by related subjects can be efficiently organized and managed for subsequent real-time analysis and mining.

    b. When an abnormality occurs in a city, decision-makers can more accurately determine the spatial range and time interval of the abnormality based on the semantic data of these trajectories. Because when an abnormality occurs, the status attributes of the relevant subject and the choices and feedback of the related parties will change.

    c. Both social media and other social business behaviors associated with these places and time periods (but not all) can be applied to analyze the causes of abnormalities.

    d. Finally, the information will be delivered to the city management department and related parties authorized by the main body in a timely manner, to quickly deal with abnormalities to avoid further state imbalance of the main body and provide a basis for future system decision-making.

2. In terms of urban data management, the data in cities are large in scale, diverse in variety, rapidly changing, and have strong spatial-temporal attributes, which are usually called spatial-temporal big data. The data management platform has two challenges: one is data management, and the other is the data center infrastructure itself.

    a. From the perspective of data management, the existing cloud platform cannot support well the management of large-scale spatial and temporal data. It is necessary to rely on spatial and temporal semantic platform technology, as shown in Fig. 4–27, to structure and associate big data to establish a dimensional vector integrating data science knowledge and industry background knowledge. This combination makes data fusion very effective in solving practical

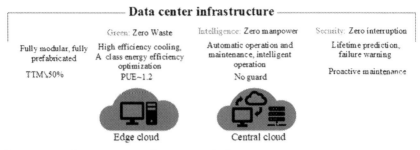

Figure 4–27 Smart city spatial semantic platform.

problems and can save a lot of data retrieval and knowledge review time to form an overall optimization of data resources.

b. From the perspective of data center infrastructure, data centers carry data storage and computing functions, while infrastructure construction cycles are long, resource consumption is high, operation and maintenance costs are high, and security challenges are high. The future smart city needs a minimal, green, smart, and secure data center to provide a digital base. At the construction and operation and maintenance level, advanced modular and intelligent technologies can be adopted. For example, the use of prefabrication and modular technology to achieve project prefabrication greatly reduces the time to go online, modular design realizes on-demand deployment, plug, and play; artificial intelligence realizes predictive maintenance of data centers and greatly improves energy efficiency.

3. Urban data analysis refers to the use of artificial intelligence technology to mine and integrate knowledge in urban big data, visually display the results, and ultimately solve industry problems. The construction of a smart city knowledge graph, as shown in Fig. 4–28, can be adopted.

In the analysis scenario of urban intelligent computing, because data scientists and industry practitioners have their own knowledge systems, there is little overlap between the two parties. Therefore industry experts must be incorporated into the data mining link through knowledge graphs to realize the intelligent integration of humans and machines. Among them, the multimode integration on the knowledge graph is the development direction of further integration.

Urban intelligent computing based on the integration of Four Networks is an effective method for smart city AI applications. It is deeply integrated with information, transportation, energy, and urban humanities

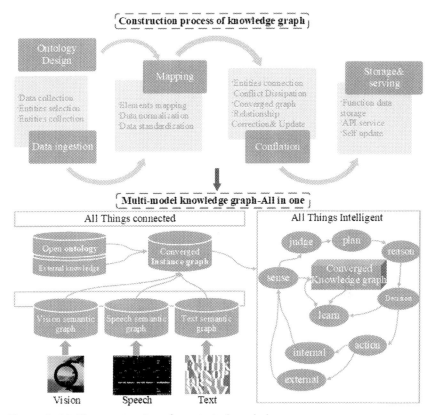

Figure 4-28 The construction of smart city knowledge.

(planning, environment, sociology, economics, and other disciplines) with the city as the background. Specifically, urban intelligent computing is a process of continuously acquiring, integrating, and analyzing multisource heterogeneous big data in the city to solve the challenges faced by the city. Urban intelligent computing combines ubiquitous perception technology, efficient data management, and powerful machine learning algorithms with knowledge graph-based intelligence and is committed to improving people's quality of life, protecting the environment, and promoting the efficiency of urban operations, helping us understand the nature of various urban phenomena, even predicting the future of cities.

References

[1] K. Zhou, S. Yang, Z. Shao, Energy internet: the business perspective, Appl. Energy 178 (2016) 212-222.

[2] W. Su, The role of customers in the us electricity market: past, present and future, Electr. J. 27 (7) (2014) 112–125.

[3] A.Q. Huang, J. Baliga, FREEDM System: Role of Power Electronics and Power Semiconductors in Developing an Energy Internet. 2009, pp. 9–12.

[4] J. Rifkin, The Third Industrial Revolution: How Lateral Power Is Transforming Energy, The Economy, and the World, Macmillan, 2011.

[5] H.M. Hussain, A. Narayanan, P.H. Nardelli, Y. Yang, What is Energy Internet? Concepts, technologies, and future directions, IEEE Access. 8 (2020) 183127–183145.

[6] W. Ejaz, M. Naeem, A. Shahid, A. Anpalagan, M. Jo, Efficient energy management for the internet of things in smart cities, IEEE Commun. Mag. 55 (1) (2017) 84–91.

[7] O. Sadeghian, A. Oshnoei, B. Mohammadi-Ivatloo, V. Vahidinasab, A. Anvari-Moghaddam, A comprehensive review on electric vehicles smart charging: solutions, strategies, technologies, and challenges, J. Energy Storage 54 (2022) 105241.

[8] X. Fang, S. Misra, G. Xue, D. Yang, Smart grid—the new and improved power grid: a survey, IEEE Commun. Surv. Tutor. 14 (4) (2011) 944–980.

[9] C.C. Chan, L. Jian, D. Tu, Smart charging of electric vehicles—integration of energy and information, IET Electr. Syst. Transp. 4 (4) (2014) 89–96.

[10] H. Lund, P.A. Ostergaard, D. Connolly, B.V. Mathiesen, Smart energy and smart energy systems, Energy 137 (2017) 556–565.

[11] H. Sun, Q. Guo, B. Zhang, W. Wu, B. Wang, X. Shen, et al., Integrated energy management system: concept, design, and demonstration in China, IEEE Electrif. Mag. 6 (2) (2018) 42–50.

[12] B. Chen, Q. Guo, G. Yin, B. Wang, Z. Pan, Y. Chen, et al., Energy-circuit-based integrated energy management system: theory, implementation, and application, Proc. IEEE 110 (12) (2022) 1897–1926.

[13] International Energy Agency—World Energy Outlook, 2021.

[14] International Energy Agency—Tracking Transport, 2021.

[15] International Energy Agency—Global EV Outlook, 2022.

[16] C. Liu, K. Chau, D. Wu, S. Gao, Opportunities and challenges of vehicle-to-home, vehicle-to-vehicle, and vehicle-to-grid technologies, Proc. IEEE 101 (11) (2013) 2409–2427.

[17] P.G. Pereirinha, M. González, I. Carrilero, D. Anseán, J. Alonso, J.C. Viera, Main trends and challenges in road transportation electrification, Transp. Res. Procedia 33 (2018) 235–242.

[18] S. Abdelhamid, H.S. Hassanein, G. Takahara, Vehicle as a mobile sensor, Procedia Comput. Sci. 34 (2014) 286–295.

[19] H. Boyes, B. Hallaq, J. Cunningham, T. Watson, The industrial internet of things (IIoT): an analysis framework, Comput. Ind. 101 (2018) 1–12.

[20] S.-W. Lin, B. Miller, J. Durand, R. Joshi, P. Didier, A. Chigani, R. Torenbeek, D. Duggal, R. Martin, G. Bleakley, Industrial internet reference architecture, Industrial Internet Consortium (IIC), Tech. Rep. (2015).

[21] A.A. Mirani, G. Velasco-Hernandez, A. Awasthi, J. Walsh, Key challenges and emerging technologies in industrial IoT architectures: a review, Sensors 22 (15) (2022) 5836.

[22] International Electrotechnical Commission – IoT 2020, Smart and secure, IoT platform, 2016.

[23] E. Sisinni, A. Saifullah, S. Han, U. Jennehag, M. Gidlund, Industrial internet of things: challenges, opportunities, and directions, IEEE Trans. Ind. Inform. 14 (11) (2018) 4724–4734.

[24] F. Sherwani, M.M. Asad, B.S.K.K. Ibrahim, Collaborative robots and industrial revolution 4.0 (IR 4.0), 2020 International Conference on Emerging Trends in Smart

Technologies (ICETST), Karachi, Pakistan (2020) 1−5. Available from: https://doi.org/10.1109/ICETST49965.2020.9080724.

[25] A. Grau, M. Indri, L.L. Bello, T. Sauter, Robots in industry: the past, present, and future of a growing collaboration with humans, IEEE Ind. Electron. Mag. 15 (1) (2020) 50−61.

[26] Z.M. Bi, C. Luo, Z. Miao, B. Zhang, W. Zhang, L. Wang, Safety assurance mechanisms of collaborative robotic systems in manufacturing, Robot. Comput. Integr. Manuf. 67 (2021) 102022.

[27] A. Schmidt, J. Reers, A. Gerhardy, Mobility as a service is accelerating in the auto industry, Accenture (2018).

[28] Knowledge graph standardization white paper. China Electronics Standardization Institute. (2019).

[29] A. Hentout, M. Aouache, A. Maoudj, I. Akli, Human−robot interaction in industrial collaborative robotics: a literature review of the decade 2008−2017, Adv. Robot. 33 (15−16) (2019) 764−799.

[30] W. Wang, X. Zhu, L. Wang, Q. Qiu, Q. Cao, Ubiquitous robotic technology for smart manufacturing system, Comput. Intell. Neurosci. 2016 (2016).

[31] R. Giffinger, C. Fertner, H. Kramar, E. Meijers, City-ranking of European medium-sized cities, Cent. Reg. Sci. Vienna UT. 9 (1) (2007) 1−12.

[32] D. Washburn, U. Sindhu, S. Balaouras, R.A. Dines, N. Hayes, L.E. Nelson, Helping CIOs understand "smart city" initiatives, Growth 17 (2) (2009) 1−17.

[33] Smart Cities Marketplace, European Commission, 2023.

[34] M. Vallianatos, Uncovering the Early History of "Big Data" and the "Smart City" in Los Angeles, 2023.

[35] T. Nam, T.A. Pardo, Conceptualizing Smart City With Dimensions of Technology, People, and Institutions. 2011, pp. 282−291.

[36] A. Hefnawy, A. Bouras, C. Cherifi, "Lifecycle Based Modeling of Smart City Ecosystem, 2020.

[37] J. Domingue, A. Galis, A. Gavras, T. Zahariadis, D. Lambert, F. Cleary, et al., The Future Internet: Future Internet Assembly 2011: Achievements and Technological Promises, Springer Nature, 2011.

[38] B. Chan, R. Paramel, The Smart City Ecosystem Framework − A Model for Planning Smart Cities, 2020.

[39] H. Azizalrahman, Towards a generic framework for smart cities, Smart Urban Development, IntechOpen, 2019, p. 3.

[40] S. Tan, J. Yang, J. Yan, Development of the low-carbon city indicator (LCCI) framework, Energy Procedia 75 (2015) 2516−2522.

[41] N. Zhou, G. He, C. Williams, D. Fridley, ELITE cities: a low-carbon eco-city evaluation tool for China, Ecol. Indic. 48 (2015) 448−456.

[42] K.-G. Kim, Methods and techniques for climate resilient and low-carbon smart city planning, Low-Carbon Smart Cities, Springer, 2018, pp. 177−213.

[43] S. Preston, M.U. Mazhar, R. Bull, Citizen engagement for co-creating low carbon smart cities: practical lessons from Nottingham City Council in the UK, Energies 13 (24) (2020) 6615.

[44] F. Starace, J.P. Tricoire, World Economic Forum − Net Zero Carbon Cities: An Integrated Approach (2021).

[45] Y. Zheng, Urban computing: driving smart cities with big data and AI, Newsl. Chin. Comput. Soc. 14 (1) (2018).

CHAPTER FIVE

Framework of Digital Renaissance with human-in-the-loop

Christopher H.T. Lee[1], C.C. Chan[2], Yaojie He[1], Chenhao Zhao[1], Huanzhi Wang[1] and George You Zhou[3]

[1]School of Electrical and Electronic Engineering, Nanyang Technological University, Singapore
[2]The University of Hong Kong, Hong Kong SAR, P.R. China
[3]National Institute of Clean and Low-Carbon Energy, Beijing, P.R. China

Abstract

The data communication between the physical entities and virtual entities creating an innovative method for industry for organizing the product lifecycle. On top of that, the humanistic knowledge and information from the customer side should be brought in, which plays a crucial role in the intelligent industry and smart society. Hence, the data, information, and knowledge interactions among the social world, physical world, and cyber world should be modeled within a dynamic system. With the human-physical-cyber system, the intelligence cognition is achieved with the ability to perform prediction, synchronization, and optimization in the digital ecology. This chapter provides the detailed elements and structure of the human-physical-cyber system, as well as the evolution, milestones, and challenges in building the human-physical-cyber system.

Keywords: Digital Twin; Hybrid Twin; Cognitive Twin; human-cyber-physical system; human intelligence

5.1 Design pattern from first principle

The integration of Four Networks and Four Flows is based on the human–cyber–physical system as shown in Fig. 5–1. The essences of human–cyber–physical system are the concepts (Digital Twin, Hybrid Twin, and Cognitive Twin) which links the physical world, cyber world, and social world, and they achieve the interactive integration among the human–information–energy–transportation network. The key in this system is the interaction of human intelligence and machine intelligence.

Conventionally, the digital industry focuses on the information and data interaction between the entities in the physical world and their counterparts in the cyber world. The virtual models assist to simulate the

Integration of Energy, Information, Transportation and Humanity.
DOI: https://doi.org/10.1016/B978-0-323-95521-8.00003-8

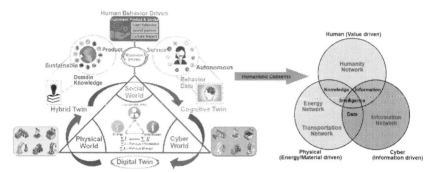

Figure 5–1 Structure of the human-cyber-physical system and Four Networks.

property, composition, and characteristics of the physical items. By utilizing the virtual models, the energy, information, and transportation networks are connected, providing the solution to design, track, and upgrade the physical entity in a dynamic and cost-effective way. However, in smart industry, the human culture is a consolidated part in the intelligent manufacturing. In such cases, it is required to digitalize the human behavior, social principle, and customized demand to be analyzed, quantified, and further utilized in guiding the digital economy. Thus the information in the social world should be interacted with the data in the physical world and information in the cyber world. Hence, the new concept named digital humanity can be formed with the human-cyber-physical system. With the help of digital humanity network and the fusion mechanism of data-knowledge and the derived intelligence, Digital Productivity will be maximized in this Fourth Industrial Revolution era.

In the human-cyber-physical system, the Digital Twin performs as the bilateral information transmission between the physical world and cyber world. Traditionally, in the cyber world, the virtual entity is established to simulate the property and characteristics of its physical counterpart. Hence, the design and optimization of such physical entity can be performed in a cost-effective, time-efficient, and ecology-friendly way. The core part of the Digital Twin comprises the initial data, modeling process, and selection criteria. For physical entities, countless data can be obtained which describes the composition, properties, and characteristics. Those data should be fully utilized in the virtual world to attain the precise virtual entities. The model in Digital Twin maps the physical item into to the cyber world, and it determines the accuracy, efficiency, and loading capability of the Digital Twin. Also, focusing on numerous design parameters and evaluation objectives, the criteria of upgrading and optimizing is essential in the Digital Twin as well. Nonetheless, the classic Digital Twin concept pays little attention to the information from social world. Also, the models for handling the dynamic data and unconstructed information are not well investigated. In such case, to integrate the information flow from the social world, other methods should be adopted.

For directly reflecting the knowledge from the social world to the physical world, a concept works as an extension of the Digital Twin that combines the data, information, and knowledge from the physical and human worlds is proposed and named as the Hybrid Twin. On the one hand, the Hybrid Twin can accurately identify the present system condition and outside environment, so that the prediction about the potential incidents and timely responses can be performed. Sufficient data and an effective physically based model can guarantee the control accuracy, deal with the system disturbances, and achieve desired performance. On the other hand, Artificial Intelligence (AI) representing the human knowledge is applied in the Hybrid Twin for data assimilation, data curation, and data-driven modeling. Therefore, as a new paradigm in simulation-based engineering sciences, the Hybrid Twin exhibits the capability of collecting the information from the cyber world and converting it to knowledge.

The Cognitive Twin is a complement of the existing Hybrid Twin with cognition ability. Since it is also based on the Digital Twin, the Cognitive Twin contains at least three parts: the physical entities, the digital representatives, and the connections between them. And beyond the Hybrid Twin, the Cognitive Twin can achieve full lifecycle management for the whole system by integrating the digital models of every part of the system in all lifecycle phases, including beginning of life (BOL), middle of life (MOL), and end of life (EOL). In addition, the Cognitive Twin has cognition abilities empowered by semantic technologies, and such ability enables it to autonomously perform human-like intelligent activities such as reasoning, prediction, decision-making, problem-solving, reaction, and so on.

5.2 Data-driven cyber-physical Digital Twin

5.2.1 Background

With the significant development in a global demanding market, commodity manufacturing becomes highly complex. Meanwhile, other than

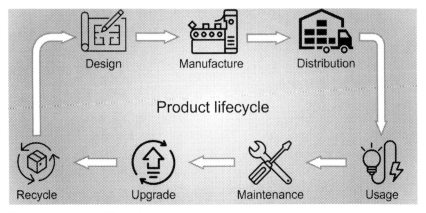

Figure 5−2 Product lifecycle phases.

the manufacture process, other phases in the product lifecycle have been attracting more and more attention in intelligent manufacturing. Hence, massive data are essential along the lifecycle of product, including design, production, distribution, usage, maintenance, upgrade and recycle, as illustrated in Fig. 5−2.

Traditionally, the data related to each phase in the product lifecycle are based on the experience of engineers. For the product design phase, several design methods have been adopted, such as Quality Function Deployment (QFD), Analytical Hierarchy Process (AHP), Theory of Inventive Problem Solving (TRIZ), Taguchi, Failure Mode, and Effective Analysis (FMEA), Kano model, Axiomatic Design and Function-Behavior-Structure (FBS) framework. However, these traditional design methods are not suitable for a product with large quantities of demanding data. The major reasons can be listed:

1. The selection for the design method, data, and information are largely dependent on the experience and relevant knowledge of designers. Generally, the designers hierarchize tremendous amount of information to improve the efficiency and quality. However, this process varies with the experience of the designers, and it is very time consuming when facing large quantities of information.

2. The information used in the traditional design methods are highly organized with certain level of clarity and consistency. During the life-cycle of a product, data can be collected from different sources, access, and forms, and it might be impractical that this is recognized and uti-lized directly by the traditional design methods.

3. The data in each phase of the product lifecycle vary from time to time. However, the promptness of the traditional design methods is not rapid enough to respond to the data due to its highly dynamic changing rate and emerging situations.

4. With the development of distributed sources and network, the data related to the customers and environment exhibit high relevance to localization characteristics. Thus the network to communicate the data from various locations and the effective mechanism to distinguish those data with tags are big challenges for the traditional design methods.

To solve the aforementioned issues and establish an effective, timesaving, and intelligent design method, the data-driven Digital Twin concept was proposed and has been attracting more and more attention. It has been listed in Gartner's top 10 strategy technology trends from 2017 to 2019 [1−3].

5.2.2 Concept of Digital Twin

Since 1960s, the twin concept has been brought in research and industry. In National Aeronautics and Space Administration (NASA)'s Apollo Program, the twin concept was realized physically by building two identical space vehicles. One of the vehicles was named the twin vehicle and stayed on earth for simulating the behavior of the space vehicle, which was sent into space. During the mission, the twin vehicle mirrored the operational condition of the space vehicle. The simulated data can be collected for determining the orders sent to the space vehicle. In those days, the simulation and data were based on the physical model and the experience of designers, and the simulation environment was built in the physical world. This twin concept was suitable for the model with fulfilled knowledge of limited working conditions, specified objectives, and oriented instructions.

The Digital Twin concept was introduced in 2002 at Michigan University by Michael Grieves in the Product Lifecycle Management (PLM). Besides the basic concept, other complementary understanding toward Digital Twin through the years are equally important. Some of them are listed in Table 5−1.

From this literature, the basic elements of the Digital Twin can be described as: physical product, virtual product, and the links between the physical product and virtual product. The physical product in the physical world is mimicked by the virtual product in the cyber world, and the links exchange information between the physical world and cyber world. Also, the Digital Twin is the model which can interact between autonomous system behavior

Table 5−1 Definitions of Digital Twin.

Authors and time	Definition
Madni et al. (2019)	A Digital Twin is a virtual instance of a physical system (twin) that is continually updated with the latter's performance, maintenance, and health status data throughout the physical system's lifecycle [4].
Wang et al. (2019)	Digital Twin can be regarded as a paradigm by means of which selected online measurements are dynamically assimilated into the simulation world, with the running simulation model guiding the real world adaptively in reverse [5].
Liu et al. (2020)	Digital Twin is a digital entity that reflects physical entity's behavioral rule and keeps updating through the whole lifecycle [6].
Singh et al. (2021)	A Digital Twin is a dynamic and self-evolving digital/virtual model or simulation of a real-life subject or object (part, machine, process, human, etc.), representing the exact state of its physical twin at any given point of time via exchanging the real-time data as well as keeping the historical data [7].
Jiang et al. (2022)	Building the Digital Twin system is a promising approach to help achieve health management, accurate life prediction, and intelligent decisions of complex physical entities throughout the whole lifecycle [8].
Hassani et al. (2022)	Digital Twin technologies include the features of being dynamic and real-time, and bi-directionally exchanging data between a physical object and its virtual representative or avatar [9].

and the environment in the physical world [10]. It means that unstructured information can be utilized in the Digital Twin without predefined criteria to preprocess the raw data from physical world. More specifically, the cyber-physical system provides the data from the lifecycle of physical product, and the data are transformed to the cyber world by edging information technology such as Big Data, Internet-of-Things (IoT), cloud computing, the fifth-generation cellular network (5G), and wireless sensor networks. Then, the information can be utilized for simulation and virtual design in the simulation and managing software (e.g., CAD, CAE, CAM, FEA, and PDM). The information generated in the cyber world is then transmitted back to the physical world, which provides the guidelines and management criteria. It is noted that the process of data and information flowing between two worlds is dynamic and simultaneous, which means the Digital Twin is capable of rapid responding to the emerging situations by real-time connection. The basic framework and elements of the Digital Twin are presented in Fig. 5−3.

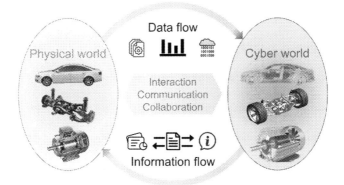

Figure 5–3 Basic framework and elements in Digital Twin.

5.2.3 Model of Digital Twin

As mentioned above, by applying the basic concept of the Digital Twin and the elements in the lifecycle of product, the model of Digital Twin can be analyzed preliminarily. The models in Digital Twin can be classified into two categories. The first focuses on describing the characteristics of a physical item (e.g., dimensions, material, properties of physical product). This type of models handles the data with explicit index and can directly determine the performance of the product. They can be employed in the process of the supply side such as manufacture, maintenance, and recycling, in which the information of a physical item is collected to be transmitted to the cyber world for analysis.

Among the design and manufacture processes, several submodels should be built in three aspects: geometry, property, and evaluation. The geometry models contain the information to manufacture the product, which defines the physical entity of the product. The property models offer the performance of the product, and it is determined by the geometry data in linear or nonlinear form. To achieve global search region and avoid predefined constraints, the high-dimension geometry models and property models are adopted in the simulation and optimization. Hence, the evaluation models play the core role in decision-making strategy (e.g., cost function and response surface method). In the simulation and optimization process, the geometry model is the input for the simulation process. The property model is the output set of the simulation and the input for the evaluation models. The evaluation models provide the data of the best candidates, or the adjustment

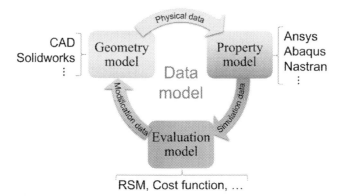

Figure 5–4 Working flow in the data models.

tendency of the data in the geometry. The relationship among the three submodels is demonstrated in Fig. 5–4.

The second type of model assembles the unconstructed information and behavior of market and customers, which can be employed in the distribution and usage parts. The raw information from market and customers (e.g., feedback, use experience, and demand) is investigated in the information model. This type of model focuses on the consumer side, and the information should be manual analysis which cannot be utilized directly. Generally, it handles the information to provide guidelines for the distribution, sale, and customer services.

The unconstructed information comes from two main sources, first the experience from the designer and secondly the concurrent information from the consumer. The major difference between these two sources is the upgrade rate. For the experience source, since it represents the consolidated and reliable knowledge coming from the practical conditions, it requires years or decades for upgrading and complementing. However, the concurrent information is collected from the processes in other phases in the lifecycle in real-time. Hence, upgrading and complementing are very frequent for this type of information source. In such a case, the information models should contain the capabilities of utilizing concrete experience and handling up-to-date information.

In the other processes, design, and upgrade, the data from the supply side and the information from the consumers' side should be considered simultaneously to evaluate the physical objective. For example, in the design part, not only the dimensions, performance, shape, and assembling

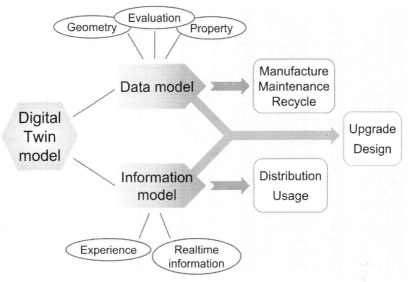

Figure 5—5 Assembling of Digital Twin model.

should be competitive to the similar products in the market, but the voice from the consumers should be considered for developing the product with specific target and population. This concept has been adopted in smart manufacturing which assesses the opinion from designers, distributors, and customers. As the result, all the possible issues related to a new product in the manufacturing and marketing chain are considered and analyzed preliminarily. In the upgrade process, the information models show higher priority because the information from the market and customers are based on the experience in use of the published products. The information from the demand side might be changing in real time. Hence, the information models in the upgrade process should exhibit the features of rapid response and high level of sensitivity. In conclusion, the focus on either data models or information models varies with different processes, as illustrated in Fig. 5—5.

5.2.4 Application case study and challenge of Digital Twin

Due to the merits of cost-saving in terms of time, resource, and financial aspects, the Digital Twin concept has been proposed in multiple disciplines, industrial areas, and research fields. For example, an Airframe Digital Twin (ADT) is built in Ref. [11] for structural components design, flight condition estimation, and damage diagnosis. The ADT contains both a data model and an information model, which can be adopted in the design process for aircraft.

Initially, the overarching ADT (OADT) works as the data model, in which the design parameters are stored and simulated. With the specific parameters search zones, objectives, and requirements, those aircraft designs which do not meet the requirements are identified and modified, which aligning with the adjustment tendency described in the evaluation model.

Afterwards, the working conditions and configurations of each aircraft are stored in the Tail number specific ADT (TADT). The initial information in each TADT is based on the experience and results from the configuration test, which provides fidelity and confidence for OADT. Along with the mission and service, the concurrent data are collected from onboard sensors, maintenance records, periodic measurement, etc. In such cases, the information in the TADT is updated and complemented in real-time, which is beneficial for evaluating the working condition of the corresponding aircraft. Focusing on various working environments and time-cost relationships, the TADT provides different operation modes to achieve economical and high-efficiency operations.

Besides in the airframe design and maintenance, the concept of Digital Twin has been widely used in other industrial fields, for example, product or equipment prognostics, health management, smart manufacturing, and engineering optimization. Those fields exhibit the features of high cost of prototyping, condition forecast, and a huge amount of unconstructed and nonlinear data. It should be noted that the Digital Twin can be applied in all phases in the product lifecycle. Thus not only the data from the physical entity are analyzed, the behavior information from market and customers are collected and utilized as well. The specific objectives are listed in Table 5−2.

Still, there are several technical challenges in the existed Digital Twin concept, these are in terms of technologies in the Digital Twin model, and the integration of the submodels with basic concept flow.

Considering the huge amount of data utilized in each submodel, and flowing through the communication link, the requirement for the hardware in dealing and transmitting the data is increasingly high. For example, the simulation for the electric machine is based on 2D form since it is identical along the machine stack. The simulation time is about several minutes to get accurate results, and the optimization time is one to two days. However, with more complex structures targeted, the simulation

Table 5—2 Application examples of Digital Twin.

Industrial field	Specific objective
Product or equipment prognostics and health management	Prediction on reengineering aircraft structural life [12]. A damage characterization method for aircraft structural health management [13]. Assessment on flight state [14]. Forecasting product health (GE).
Smart manufacturing	Digital Twin shop-floor [13,15]. Industrial production line [16]. Predictions about properties and serviceability of components [17]. Geometry assurance in individualized production [18]. Improvement on manufacturing efficiency and flexibility (DXC).
Engineering optimization	Optimizing equipment operation [10]. Optimizing the behaviors of complex system [19].

turns into 3D model. As a result, the simulation time is increased to several hours. In this case, the amount of data in the property model becomes very large. Meanwhile, due to the consideration of multiple physical fields, the objectives to evaluate the design from the property model are expanded. Hence, the amount of data in the evaluation model and the information flow between property model and evaluation model are increased as well. The huge data burden leads to a great challenge to the processing and transmitting capabilities of the hardware.

Toward the information model, the data challenges are in other aspects. Firstly, the information comes from large sources and is collected in different forms. It varies from specific statistics to fuzzy descriptions. Some of the information is difficult to quantify. Thus the models which can directly utilize the types of information in the product lifecycle require further investigation. Also, the integration of different types of information model to form a big picture is a popular topic in interdisciplinary fields. Finally, since the Digital Twin is updated with concurrent information, how to rapidly respond to a real-time event and make accurate predictions are opportunities in the future development of the Digital Twin.

In summary, the concept of the Digital Twin provides a novel method toward PLM field. It establishes a virtual item in the cyber world which reflects the property, composition, and characteristics of the counterpart in the physical world. In the cyber world, the state and performance can be simulated to provide the guidelines and criteria for handling the corresponding physical entity. Moreover, with techniques such as Big Data and IoT for collecting information in the cyber world and physical world, respectively, the Digital Twin is able to track and update during the lifecycle of the product in real-time. Consequently, the data-driven cyberphysical Digital Twin is a promising tool in the intelligent manufacture industry.

5.3 Knowledge-driven human-cyber Hybrid Twin

5.3.1 Background

As mentioned in the last section, by using computational models and measuring the function of time and interest, the information model in the Digital Twin converts various data from practical physical system into information. However, there still exist some challenges in the information model system. For example, this modeling technology is not only required to model simple objects but also to convert the data of complex objects relationships (which may contain unstructured data) accurately and effectively.

In the middle of the 20th century, with the development and wide applications of information technologies such as computer, communication, and digital control, the industry entered the era of digitalization, which significantly facilitated the development of information model technology. The information revolution marked by digitization leads and promotes the third industrial revolution. As Fig. 5—6 indicates, the

Boost productivity & quality
- Accelerate important processes
- Minimize product errors

Enhance business resilience
- Control risk of planning scenarios
- Improve collaboration

Facilitate sustainable innovation
- Reduce energy consumption
- Provide usage date

Figure 5—6 Benefits of digitalization.

automation degree, efficiency, quality, and stability of each phase in the product lifecycle and the ability to solve complex problems have been markedly enhanced. Also, the dissemination, utilization, and inheritance efficiency of knowledge are effectively improved [20].

However, the analytical solutions of models used in science and engineering are frequently affected when they become overly complex. Moreover, since it is very time-consuming to process the complex models directly, the simplification toward complex mathematical objects is crucial in boosting the processing efficiency. Hence, the discrete systems processed by digital computers have attracted more and more attention since they cost less computing time and need fewer sampling points for model solutions computation.

Although data-driven research has always been prevalent, most of scientific fields had a large data explosion toward the end of the 20th century, notably in the field of engineering. With the development of the Digital Twin, which is a data-driven representation of the reality, data have been widely engaged into the best practices in various fields where models for physical entities are defective or undependable. Huge volumes of data were categorized, displayed (despite the multidimensional nature), curated, and analyzed by using advanced machine learning algorithms. In addition, to establish the links between certain inputs and specific outputs, some methodologies that were presumed competent to explain and infer the outputs were presented. They use a range of AI techniques, such as neural networks, decision trees, random forests, and linear and nonlinear regressions. Hence, they are the so-called "model learners." However, one of the potential problems of the Digital Twin which should be considered is that the volume of transmitted data can be overwhelming, but the practical use of this deluge is not obvious. Is there any method to put this data to good use? Can human knowledge be applied in the Digital Twin?

5.3.2 Concept and challenges of Virtual Twin

Before combining human knowledge with the Digital Twin, the concept of Virtual Twin should first be introduced. The Virtual Twin is a simulation-based perfect replica of the product or other objects. As Fig. 5−7 illustrates, it uses mathematical models to simulate a physical system and describe its complicated behavior, this plays a significant role in the preceding industrial revolution. Here, physics-based simulation models contain the behavior of products and systems using analytical and

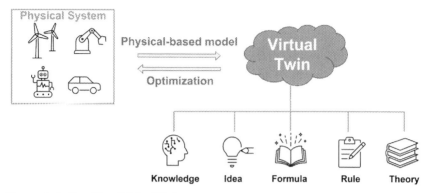

Figure 5−7 Benefits of Virtual Twin.

numerical techniques [21]. One example is the use of Finite Element Analysis (FEA) for structure simulation. A physics-based simulation of reality is what the Virtual Twin represents, and the firm foundations provided by centuries of verified physics are its greatest asset.

Generally, the analytical method, which is the core of Virtual Twin, exhibits the characteristic of faster computation speed. However, its main drawback is the computation accuracy. With the development of science and technology, the system become much larger and more complicated than before, and there are more reactions and interactions among different parts of the whole system. Therefore some of the previous domain knowledge regarding physical rules or mathematical models may not be effective and accurate in the practical situations. In addition, to balance the relationship between the computation speed and simulation accuracy, the system complexity is reduced by simplifying the system nonlinearity and approximating some component in mathematical models. All these elements deteriorate the accuracy of physical-based simulation.

Nowadays, the numerical simulation has been used in various disciplines. This method significantly enhances the accuracy of simulations by enabling precise designs and virtual evaluation of system reactions. Engineering practice typically uses static numerical models. The major types of static numerical models—finite element, finite volume, and finite difference—are frequently used in the creation of intricate engineering systems and their subsystems. These models are static because they are not anticipated to be continuously provided with data. However, the strict real-time requirements of some projects or tasks, especially for control purposes, cannot be accommodated by the characteristic time of standard engineering simulation methodologies. It is a common practice to apply methods based on the usage of ad hoc—or "black box"—models of the system to ensure real-time simulation for control (in the sense that they relate some inputs to some outputs, encapsulated into a transfer function) [22]. This modified system enables real-time operation. However, when compared to the simulations with high-fidelity, such as Finite Element techniques, it becomes too coarse.

Therefore considering the limitations of numerical simulation approach, it is necessary to apply data-based dynamic models to substitute the original static method. Digital Twin with massively gathered and properly curated data gives interpretational keys to an impending event. This enables real-time decision-making and improves data-based predictive maintenance as well as efficient inspection and management in the product lifecycle. The Digital Twin has become exponentially popular because of the documented accomplishments and the great potential for development. Data-driven models enable the representation of systems in all their complexity and ensure real-time access to any necessary data. However, just as the development of verified models required time and effort during the previous engineering revolution, reaching an optimal learning stage also takes considerable time. In addition, the physical-based solutions derived by human knowledge have been very well established and widely tested in the last century. Replacing the rich history of engineering sciences, which have shown their potential with spectacular achievement over the course of more than a century, results in emotions of resentment and a waste of learned knowledge [23].

Considering the advantages and drawbacks of both Virtual Twin and Digital Twin, it is naturally assumed that if these two twins can be combined, they can play a larger role in the practical applications. It means that the human knowledge and guidance representing the virtual twin can be applied to the cyber world achieved by the digital twin to help solve both of their problems. Therefore the concept of Hybrid Twin is proposed.

5.3.3 Concept and model of Hybrid Twin

In the earlier development of engineering modeling, one of the schemes proposed that the Virtual Twin is used offline during the design phase and is replaced by data-based digital counterparts during online operation. Even if this plan improves the response rate in a certain degree, its scope of application is limited. The most appealing approach appears to be a combination of the Virtual Twin and Digital Twin. Nevertheless, before integrating these two types of twin concepts, the real-time restriction of physical models must be firstly resolved. Adopting powerful computers or contemporary supercomputing facilities is a worthwhile option. However, this method cannot eliminate all the issues mentioned, and it severely restricts access to the necessary simulation infrastructure.

The human knowledge including applied mechanics, theoretical and applied mathematics, and computer science made significant contributions to modeling and simulation techniques at the end of the 20th and beginning of the 21st centuries. One of the key accomplishments is the development of model order reduction (MOR) approaches [24]. This method does not simplify the model; instead, it keeps using models with trusted and well-established explanations of the relevant physics. To accommodate real-time limitations, it relies on an acceptable approximation of the solution that enables the simplification of the solution process without compromising the accuracy of the model solution. The MOR-based discretization approaches significantly reduce the computation time. Consequently, it enables effective simulation, optimization, inverse analysis, and simulation-based control to be carried out while being constrained by real-time requirements.

Then the next generation of twin concept called "Hybrid Twin" was born by combining physically based models within a MOR framework (for accommodating real-time feedback) and data-science, which integrates the Virtual Twin with the Digital Twin (shown in Fig. 5—8) [25].

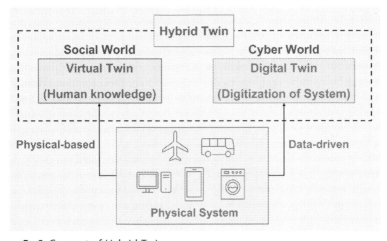

Figure 5—8 Concept of Hybrid Twin.

It is an extension of the Digital Twin that associates the separate Digital Twin models to identify, predict, and convey less-than-ideal (but predictable) behavior of the physical counterpart before it happens. More importantly, with the guidance and effect of human knowledge, the hybrid twin can optimize, monitor, and regulate the behavior of the physical system while integrating data from multiple sources (such as sensors, databases, and simulations) to gain stronger predictive capabilities.

To be specific, high-fidelity models were initially explored virtually in real-time situations using conventional computing platforms. Once they are linked with data-driven application systems, significant differences between the expected and observed responses are detected. These discrepancies between reference and observations can be traced back to errors in the models used, parameters, or temporal evolution. By creating instantaneous data-driven models, this could eventually fix the discrepancy between the data and model reference. Now with the real-time data combined with the simulation model, the simulation accuracy can be increased by minimizing the error. Therefore the Hybrid Twin is capable of accurately identifying the present system condition and outside environment, so that the potential accidents and incidents can be predicted, and rapid responses can be outperformed. Moreover, the real-time control and decision-making can be efficiently conducted by the Hybrid Twin. Sufficient data and effective physically based models ensure the accuracy of control strategy, deal with the system inside and outside disturbances, and make decisions for various instructions and emergencies.

On the other hand, as an important part of human knowledge integration, the AI techniques are applied in the Hybrid Twin to drive the industry progress. The modeling ability of each phase of a product lifecycle is fundamentally improved, which effectively optimizes the system operation. In addition, as Fig. 5—9 indicates, from the large volume of information provided by the cyber system, AI techniques make the Hybrid Twin able to generate knowledge based on existing and collected information and produce better use of the knowledge. As a result, the

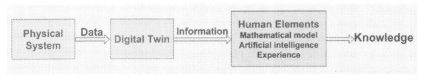

Figure 5—9 How the knowledge is generated by Hybrid Twin.

production, utilization, inheritance, and accumulation efficiency of knowledge are revolutionized, and the marginal productivity of knowledge, as a core element, is significantly improved. Moreover, this stage lays the foundation for the generation of machine intelligence. The process of product design, quality control, production management, and other links and integrations are digitalized and cyberized.

5.3.4 Application of Hybrid Twin

5.3.4.1 Electric vehicle

By applying the cutting-edge Hybrid Twin concept, the comfort and safety of electric vehicles (EV) are improved [26]. Once this twin system is integrated into an EV via a clever human—machine interface, it will enable the driver to get real-time alerts if the car anticipates an issue during an intended journey. Also, the interface provides the driver with advice on how to optimize the in-car comfort over a long driving distance. Furthermore, under various traffic conditions and weather predictions, the interface can offer an optimized driving route, making sure the driver and the passengers get rid of hazardous driving circumstances together with a highly efficient driving process.

In addition, energy management to optimize the vehicle driving distance is one of the essential topics for EVs. In terms of battery capacity, it is a big challenge to balance the comfort requirements with operation distance and time. This problem can be relieved by using Hybrid Twin technology, which is fused with the knowledge of human behavior, material physics, and 3D simulation approach. To evaluate the thermal comfort, this 3D model includes detailed models of the cabin, smart seat, and human needs. It engages with the system of the entire vehicle using a comprehensive, occupant-focused strategy. Within the MOR frame, the immediate response to intricately interwoven needs for comfort, safety, and driving range under various working situations is enabled. The combination of large amount of real-time data with expertise knowledge leads to outstanding outcomes: the driving range of electrical vehicles was increased by up to 40% while providing the greatest levels of cabin safety and comfort.

5.3.4.2 WindTwin project

A Digital Twin-based system is constrained by the quantity and distribution of available sensors [27]. For instance, a typical wind turbine adopts six to eight strain sensors per blade and one or two temperature sensors in the gearbox. Those sensors offer a continuous stream of data for monitoring the working conditions. The data can also be used to create, validate, and improve the system models. However, wind farm managers are not only interested in the mechanical and thermal behavior from these sensors. They also want a tool that enables detection on the potential measures to prevent operation failure or poor performance, as well as the ability to foresee all the possible outcomes when the sensor starts sending an unusual pattern of data. Such tool to suggest the optimum course of action can reduce both the overall cost and the operational disruption rate.

By the end of 2016, there were about 250,000 wind turbines worldwide. And it is estimated that the market for operation and maintenance is around 20 billion USD at the end of 2022. The annual maintenance cost is typically 5% of the capital cost, and it can occasionally be as high as 10%. Hence, for wind farm operators, the maintenance phase should be improved to make it higher cost-effectiveness. The WindTwin project initiatively intends to simplify the monitoring and maintenance procedures. The objective is to boost the accessibility and dependability of wind farms, so that the requirements of both onshore and offshore farms are satisfied. All pertinent physics knowledge at the subsystem level are included in a specialized Hybrid Twin-based solution, accompanied by adequate models. Instead of using predetermined and schedule-based techniques, operators can use the WindTwin to identify operational abnormality and implement condition-based maintenance. By allowing operators to virtually test maintenance improvements prior to deployment, Hybrid Twin models surely reduce downtime and save maintenance expenses. Additionally, they can provide more precise control of wind turbines to maximize production and improve performance.

The Hybrid Twin is made up of three main components:

1. A simulation core that can handle advanced AI techniques for data assimilation, data curation, and data-driven modeling.
2. Advanced strategies that can solve complex mathematical problems representing physical models under real-time constraints.
3. A mechanism for online adapting the model to changing environments.

These three functions are incorporated into the Hybrid Twin as a new paradigm in simulation-based engineering sciences. Since the Hybrid Twin is an extension of the Digital Twin, some of the problems in Digital Twin have been solved. For example, the isolated Digital Twin can cooperate, connect, and communicate with other twins within the frame of Hybrid Twin. Moreover, it is able to predict potential accidents and events based on changes in the internal system states and external environment, thereby enabling quick responses to real-time events.

At present, by introducing the human and social elements, the Hybrid Twin can generate the knowledge from the cyber world. However, the Hybrid Twin can still be improved with the development of science and technology. For instance, since it lacks the cooperation ability along the production lifecycle, the hybrid twin can only play roles in a single phase or single system. It is expected that the model system has its own intelligence to integrate the whole system and deal with all the incidents during operation. Thus the intelligence-driven model will enable the solution of more complicated problems and enable the optimization of the system behavior.

5.4 Intelligence-driven human-cyber-physical Cognitive Twin

5.4.1 Background

In the previous sections, the concepts of the Digital and Hybrid Twins have been introduced in detail. As described, physical, cyber, and social worlds are bridged by the Digital and Hybrid Twins, hence enabling the generation of dynamic data and static knowledge from the physical entities. With these data and knowledge, basically three functions are empowered as shown in Fig. 5—10. Firstly, the physical entities are connected and configured which enables the obtainment of data. Then, based on the collected data, monitoring and visualization can be conducted. Furthermore, with model-based and data-driven approaches,

Connect & Configure	Monitor & Visualize	Analyze & Predict	Cognitive
• Physical assets, IIoT, etc. • Connect data sources • Collect, aggregate, explore	• Data visualization • Monitoring platform • Statistical analysis	• Model-based analysis • Machine/Deep learning • Prediction/recommendation	• Complex and unpredicted behaviors • Sensing and reasoning. • Decision making, etc.

Figure 5—10 Cognitive engineering journey [28].

system analysis and predicting are also available. These three functions support the typical services among several twins, including process monitoring, abnormality detection, and predictive maintenance.

However, since no machine intelligence has been created, these three functions are only able to deal with known issues, and human decision-making is required. Thus for the fourth function in Fig. 5−10, extra abilities are demanded, such as dealing with unknown or unpredictable behavior in the application of sensing and reasoning for autonomous decision-making. Besides, it is a challenge to integrate the twin models among each lifecycle phases in complex industrial systems. Also, a variety of stakeholders may develop different models based on their own standards, experience, and criteria. Hence, the data structures of those models are typically heterogeneous in terms of syntax, schema, and semantics.

To achieve higher automation and full lifecycle phases management, the concept of Cognitive Twin has been recently proposed. Based on the concept of Digital Twin, Cognitive Twin also contains the physical entities in physical world, their digital or virtual representations in cyber world, and linkages between them. Since a complex industrial system has numerous digital models of its components which contains their unique status along the system's full lifecycle, Cognitive Twin aims to integrate all the digital models with unified semantic topology. Hence, the existing function of the Digital and Hybrid Twins are improved, and more services can be provided. In general, Cognitive Twin is an expansion of current Digital Twin and Hybrid Twin with communication, analysis, and intelligence capabilities in three layers: access, analytics, and cognition [29]. It exhibits the ability to pinpoint the dynamics of virtual model evolution, understand how virtual models interact with each other, and improve the autonomous decision-making. The relationship among Cognitive Twin, Hybrid Twin, and Digital Twin is shown in Fig. 5−11.

5.4.2 Characteristics and structure of Cognitive Twin

Based on the related works, the advanced characteristics of the Cognitive Twin can be concluded in three parts, namely, cognition capability, overall system lifecycle management, and autonomy capability.

5.4.2.1 Cognition capability

Literally, the ability to conduct cognitive functions (e.g., attention, perception, comprehension, memory, reasoning, prediction, decision-making, problem-solving, reaction) is the most crucial trait that a Cognitive Twin should possess.

Cognitive Twin		
Multiple system hierarchy levels	Cognition capability	Multiple lifecycle phases
Hybrid Twin		
Domain knowledge (Social world)		
Digital Twin		
Virtual entities (Cyber world)	Linkage between cyber world and physical world	Physical entities (Physical world)

Figure 5−11 Relationship among Digital Twin, Hybrid Twin, and Cognitive Twin.

As a result, a Cognitive Twin is described as a twin model having the ability to dynamically recognize complicated and unpredictable actions. And it is enabled by the rapid advancement of semantic technologies including AI and ubiquitous sensing technologies.

5.4.2.2 Overall system lifecycle management
A Cognitive Twin should provide the interlink among the digital models in the entire system, which can be categorized by being based on different lifecycle phases. The classified models can be arranged in a hierarchy system as shown in Fig. 5−12. The system's complete lifecycle phases comprise the Beginning of Life (BOL), which refers to the stage before the product put into use including design, manufacture, and distribution; Middle of Life (MOL), which refers to the stage of usage and maintenance; and End of Life (EOL), which is the stage for postusage including upgrading and recycling.

5.4.2.3 Autonomy capability
A Cognitive Twin should be capable of handling autonomous tasks with no or only a minimal amount of preliminary manual participation. The autonomy capability of a Cognitive Twin is derived from and enhanced by its cognition ability. For instance, based on the results of perception and prediction, a Cognitive Twin can act independently, making decisions for design, manufacture, operations, and so on.

In general, the Cognitive Twin can be described as a digital representation of a physical system that has been enhanced with specific cognitive

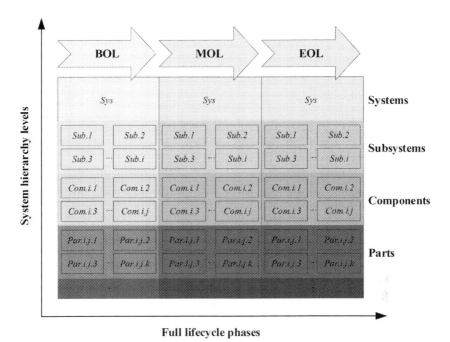

Full lifecycle phases

Figure 5–12 System structure diagram dividing in system hierarchy levels and life-cycle phases.

abilities and support to carry out autonomous activities. It also includes several digital models that are semantically linked to each part in different lifecycle phases of the physical system, and it is continuously developed alongside the entire lifecycle of the physical system. The structure of Cognitive Twin is shown in Fig. 5–13, where a top-level ontology is utilized to combine various ontologies synchronized with virtual entities throughout the lifecycle phases to assist reasoning for cognitive decision-making.

5.4.3 Enabling technology for model management

As mentioned before, it is challenging for different twin models to be kept synchronous and aligned since they incorporate heterogeneous data, information, and expertise from different stakeholders and information sources. Hence, semantic technologies are regarded as the foundation of the Cognitive Twin because of their benefits as enhancing data interoperability and developing cognition capacities. Among the semantic technologies, ontology and the knowledge graph are two of the most promising solutions.

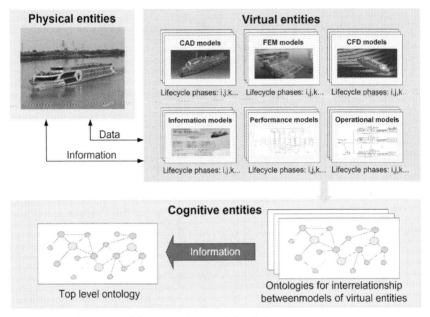

Figure 5–13 Structure of Cognitive Digital Twins [29].

5.4.3.1 Ontology

Ontology is regarded as one of the feasible foundation methods for cognitive systems because it can formalize the ontological characteristics of physical phenomena in a way that is consistent with common sense. A typical technique is to extract information from unstructured data using a knowledge graph, incorparate the information into an ontology, and then use a reasoner to build new knowledge from that ontology.

Various established approaches, tools, and languages are available today to facilitate ontology construction. For different application settings, multiple ontologies have been developed using numerous languages and techniques. However, it is a difficult undertaking to integrate all the ontologies into a single framework to ensure their interoperability. As a result, this is a prevalent issue when creating Cognitive Twin models for intricate systems. One of the solutions is to use a hierarchical framework to combine the application ontologies under a single top-level ontology that incorporates a collection of general vocabularies regularly used across all domains. These vocabularies follow a certain methodology and are appropriately structured and specified. The top-level ontology serves as a base for generating more in-depth applications and domain-specific ontologies at a lower level. And the adoption of the top-level ontology assures semantic interoperability among those lower-level ontologies.

5.4.3.2 Knowledge graph

The most potential enabler for Cognitive Twin vision is the knowledge graph. It describes the topologies of both organized and unstructured data using a graph model made up of nodes and edges, enabling the formal, semantic, and structured representation of information. In a knowledge graph, nodes are used to represent entities or literalized raw values, while connecting edges indicate the semantic relationships between nodes. By creating semantic links across disparate data sources, knowledge graphs make it possible to extract and proceed the underlying knowledge from them. Consequently, for the Cognitive Twin vision, the incorporation among all twin models is achievable.

Therefore based on the concepts of ontology and knowledge graph, static knowledge such as design manuals and operating experience can be combined with dynamic operating data to generate dynamic knowledge. While dynamic knowledge can be learned, acquired, and generated, it will be used at the system level. The iterative process of the gradually evolving knowledge computing platform provides a scientific guarantee for the system's intelligent decision-making.

5.4.4 Function layers of Cognitive Twin

To introduce the Cognitive Twin functions in detail, a six-layer progression model is formed in Fig. 5—14. It can be seen from the figure that, there are four function layers sandwiched by physical entities and user interaction. Those four function layers are responsible for management in all aspects including data, models, services, and twins.

5.4.4.1 Data management

Different forms of data and information are collected from individual sources in this layer, which reflects the status of various physical system hierarchy levels and system lifecycle phases. Following that, the data are transmitted via specific protocols to the data and metadata repositories. Specific brokers and adapters enable the data ingestion method. The data can be processed and analyzed utilizing data mining techniques enabled by edge, fog or cloud computing depending on the specifications for the relevant services.

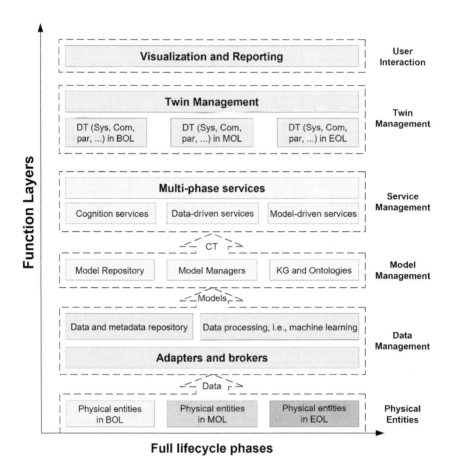

Figure 5–14 Six-layer progression model of Cognitive Twin.

5.4.4.2 Model management

To store and maintain various twin models (e.g., first-principle models, empirical models, and knowledge-driven models), the semantic technologies, such as knowledge graph and ontology are used in the model management layer. To further manage the models, a model manager is adopted which provides top-level services including:

1. Monitoring and managing modifications toward a group of connected models.
2. Planning, organizing, and coordinating all models and their traceability-related procedures.
3. Identifying and fixing consistency violations to make sure the models are semantically and syntactically correct.

4. Developing models in collaboration with stakeholders throughout full lifecycle to ensure that their needs are fulfilled.
5. Employing a range of techniques to identify, model, analyze, measure, enhance, optimize, and automate modeling processes by stakeholders.
6. Supporting whole system full lifecycle phases management.

5.4.4.3 Service management

At the service management layer, various business services, including data-driven, model-driven, and cognition services are registered and orchestrated in accordance with the needs of the application scenarios. In addition, data-driven and model-driven services concentrating on both single and multiple lifecycle phases are also offered at the service management layer.

5.4.4.4 Twin management

A Cognitive Twin involves numerous twin models representing various subsystems, components, and segments. These twin models all undergo lifecycle phase modifications. Hence, through the whole lifecycle, the twin management layer intends to assist to renew and coordinate the twin models.

5.4.5 Applications and challenges of Cognitive Twin

The Cognitive Twin is a novel concept that has yet been widely used or tested in business. Most published works are focusing on either examining the theoretical angles of Cognitive Twin or constructing the Cognitive Twin vision. Nevertheless, several ongoing studies and projects strive to confirm the Cognitive Twin's viability by using it in various industry contexts (Table 5−3).

Table 5−3 Cognitive Twin applications.

Projects	Aims
COGNITWIN [30]	Improve the cognitive capacities of the current process control systems, enabling self-organization, and providing answers to unexpected behaviors.
FACTLOG [31]	Merge data-driven and model-driven Digital Twins to enhance complicated process systems' cognitive capacities.
QU4LITY	Assist autonomous quality in order to deliver open, standardized, and transformable zero-defect manufacturing products, services and models.

To completely accomplish the Cognitive Twin vision, numerous challenges need to be overcome. Although some of the issues can be mitigated by the enabling technologies, working toward other aspects is required in the future, especially in the knowledge management area, Digital Twin model integration, and standardization and implementation.

In conclusion, the Cognitive Twin is an advanced twin model concept with cognition ability. Different from data-driven Digital Twin and knowledge-driven Hybrid Twin, the Cognitive Twin is driven by intelligence generated by semantic technologies. With the machine intelligence, Cognitive Twin can diagnose unpredictable faults and make decisions automatically in the same way as human beings. Since the Cognitive Twin links the cyber, physical, and the social worlds together, it provides both machine intelligence and human intelligence in the physical-cyber-human system.

5.4.6 Future milestones

For achieving Digital Renaissance with the consideration of humanistic data, the human–cyber–physical system is introduced, which shows both human intelligence and machine intelligence. Soon, several foreseeable milestones can be brought out which provides the guidelines and blueprint for Digital Ecology including both digital economy and digital humanity. The data, knowledge, and information flow in Digital, Hybrid, and Cognitive Twins are illustrated in Fig. 5–15.

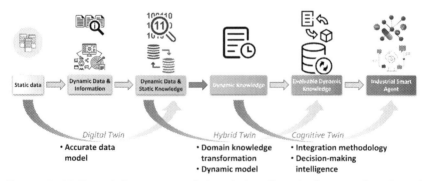

Figure 5–15 Data, information, and knowledge flow in human-cyber-physical system.

5.4.6.1 Accurate data model in Digital Twin

In the Digital Twin, the static data of products are mapped into cyber world for simulation and optimization. With the help of information preliminarily defined by designers, the static data is transformed into dynamic data which is evolved by itself. After this process, the static knowledge of such product can be obtained, which can be utilized into the manufacture phase, and further provide the basic benchmark for the upgrading phase. The accuracy and speed of data processes in the Digital Twin significantly rely on the data model. The virtual entity description, multiphysics simulation, and evaluation in the cyber world are based on the data model. Generally, in the finite element method, the simulation speed and accuracy are traded off to achieve both high efficiency and accuracy. However, to obtain an acceptable accuracy, the minimum processing time is largely dependent on the topology of the data model. Consequently, for establishing the basis for Digital Twin and human–cyber–physical system, the development of the data model should be prioritized.

5.4.6.2 Directly transformation of domain knowledge into cyber world

In the intelligent human–cyber–physical system, the information from human beings such as experience, standards, and specifications should be integrated within the analysis in the cyber world. This is a fundamental step in which the domain knowledge is transmitted to the cyber world. Typically, the humanistic information is predefined by the designer ahead of the manufacture of product, especially in the design phase of the product lifecycle. However, in intelligent manufacture, with higher integration and complex level, the industrial design requires interdisciplinary knowledge. In such case, the digitalization toward domain knowledge is imperative. By this process, the domain knowledge toward different phases in the product lifecycle can be utilized in real time. The human resource can be reduced, and the domain knowledge can be more efficiently utilized with less time consumption and human resource. Hence, the methods for handling the domain knowledge need further investigation and application in the virtual twin part of Hybrid Twin.

5.4.6.3 Dynamic models for processing and transmitting real-time information

Another function of Hybrid Twin is to handle the dynamic data from the social world. It should be noted that the timeliness of the virtual twin is essential since the concurrent human behavior varies rapidly over time. Also, massive amount of data is obtained when the dynamic human behavior information is digitalized. Hence, the capability of accurate digitalization, data processing, and rapid response to real time emergencies are the essential targets for the development of human-in-loop Hybrid Twin. Along with the dynamic data from social world and static knowledge obtained from the Digital Twin, dynamic knowledge can be formed and implemented, amended, and utilized online. Afterwards, the dynamic knowledge can be evolved by itself when it is applied in a new working condition, the targeted customers is switched, or the market policy is changed. Consequently, for each phase in the product lifecycle, the information of humanistic network is included, and the basic function of machine intelligence is achieved.

5.4.6.4 Integration methodology along the product lifecycle

In the Cognitive Twin, the self-evolved dynamic knowledge generated by the Hybrid Twin in each product lifecycle phase should be integrated. In other words, the product state is tracked and simulated by using such integration methodology. It not only provides the information for developing the product strategy but also predicts the consequences under various working conditions, sale strategies, and upgrading plans. In another aspect, under single working conditions containing numerous products, the effect of each product and the mutual effect among those products can be simulated by this methodology. It means that integration methodology can be both scenario-oriented and product-oriented. It composes a network mapping the real-life situation into the cyber world.

Also, by utilizing the integration methodology, the product evaluation is expanded from single phase to a big picture, which is high-level machine intelligence to be achieved in the Cognitive Twin. During this process, the information from the humanistic network should be included. Traditionally, the opinions from the consumer side are collected in postmanufacture progress. However, with the dynamic knowledge from various products in different lifecycle phases, the information from consumer side can be obtained initially. With evolved dynamic humanistic knowledge in each phase from Hybrid Twin after the manufacture, the initial consumer information is amended dynamically and utilized for evaluating the corresponding product, providing initial consumer information for other products and establishing the knowledge base for industrial intelligence.

5.4.6.5 Fully autonomous and intelligent decision-making smart agent

To achieve a fully intelligent digital industry, the automatic decision-making system and strategy composition scheme are required. Hence, the human—machine smart agent is put forward which achieves "perceive-think-execute-evolve." The smart agent consists of four intelligent layers, namely: interaction, connection, hub, and application. The core purpose of the intelligent hub is the enablement of data, AI, development, and integration.

Human—machine smart agent has four major characteristics:

1. Hierarchical decoupling: overall planning, step-by-step implementation, open cooperation, connection between customers and ecology, and clear definitions on business scope and capability boundaries.
2. Multiple and heterogeneous cloud: providing a flexible multivendor strategy, users can choose the appropriate platform according to the application scenarios.
3. Hybrid computing power: support information innovation to solve the troublesome problem, help customers achieve national independence and controllable goals based on customer business strategies.
4. Performance indicators: provide products and services with different performance indicators on demand and assist in the formulation of industry standards.

In the integrated human—machine smart agent, data, information, and knowledge are spontaneously flown in the human-cyber-physical system to achieve "perceive-think-execute-evolve" process. Furthermore, the dynamic knowledge can be evolved by the system itself, which provides the additional information for human to explore the undiscovered area or underinvestigated solutions toward industrial breakthrough. Consequently, in the digital economy, the fully intelligent ecology is attained.

With the humanistic network, the urban industrial interconnection between physical world and cyber world is expanded into a higher level of interaction. The humanistic knowledge works as a novel criterion in the PLM, in both static and dynamic ways. By establishing real-time, end-to-end, multidirectional communication and data sharing between people, products, systems, assets, and machines, each product and production process can be independently monitored to perceive and understand the surrounding environment. The humanity network, the fusion mechanism of data-knowledge, and the derived intelligence can help in maximizing Digital Productivity in this Fourth Industrial Revolution era.

References

[1] Gartner's Top, 10 Strategic Technology Trends for 2017, 2019. https://www.gartner.com/smarterwithgartner/gartners-top-10-technology-trends-2017.

[2] Gartner's Top, 10 Strategic Technology Trends for 2018, 2019. https://www.gartner.com/smarterwithgartner/gartners-top-10-technology-trends-2018.

[3] Gartner's Top, 10 Strategic Technology Trends for 2019, 2019. https://www.gartner.com/smarterwithgartner/gartners-top-10-technology-trends-2019.

[4] Madni A., Madni C., Lucero S. Leveraging digital twin technology in model-based systems engineering. Systems, 2019.

[5] P. Wang, M. Yang, Y. Peng, J. Zhu, R. Ju, Q. Yin, Sensor control in anti-submarine warfare—a digital twin and random finite sets based approach, Entropy 21 (2019).

[6] M. Liu, S. Fang, H. Dong, C. Xu, Review of digital twin about concepts, technologies, and industrial applications, J. Manuf. Syst. (2020).

[7] M. Singh, E. Fuenmayor, E.P. Hinchy, Y. Qiao, N. Murray, D. Devine, Digital Twin: origin to future, Appl. Syst. Innov. (2021).

[8] Z. Jiang, H. Lv, Y. Li, et al., A novel application architecture of digital twin in smart grid, J. Ambient. Intell. Hum. Comput. (2022).

[9] H. Hassani, X. Huang, S. MacFeely, Impactful Digital Twin in the healthcare revolution, Big Data Cognit. Comput. (2022).

[10] R. Rosen, G.V. Wichert, G. Lo, K.D. Bettenhausen, About the importance of autonomy and digital twins for the future of manufacturing, IFAC-PapersOnLine 48 (3) (2015) 567−572.

[11] E. Tuegel, The airframe digital twin: some challenges to realization. 53rd AIAA/ASME/ASCE/AHS/ASC Structures, Structural Dynamics and Materials Conference 20th AIAA/ASME/AHS Adaptive Structures Conference 14th AIAA, 2012, 1812.

[12] E.J. Tuegel, A.R. Ingraffea, T.G. Eason, S.M. Spottswood, Reengineering aircraft structural life prediction using a Digital Twin, Int. J. Aerosp. Eng. (2011) 1−14.

[13] F. Tao, J. Cheng, Q. Qi, M. Zhang, H. Zhang, F. Sui, Digital Twin-driven product design, manufacturing and service with big data, Int. J. Adv. Manuf. Technol. (2017) 1−14.

[14] B.T. Gockel, A.W. Tudor, M.D. Brandyberry, R.C. Penmetsa, E.J. Tuegel, Challenges with structural life forecasting using realistic mission profiles. 53rd AIAA/ASME/ASCE/AHS/ASC Structures, Structural Dynamics and Materials Conference 20th AIAA/ASME/AHS Adaptive Structures Conference 14th AIAA, 2012.

[15] F. Tao, M. Zhang, J. Cheng, Q. Qi, Digital Twin workshop: a new paradigm for future workshop, Computer Integr. Manuf. Syst. 23 (1) (2017) 1−9.

[16] J. Vachálek, L. Bartalský, O. Rovný, D. Šišmišová, M. Morháč, M. Lokšík, The Digital Twin of an industrial production line within the industry 4.0 concept. IEEE International Conference on Process Control, 2017, pp. 258−262.

[17] G.L. Knapp, T. Mukherjee, J.S. Zuback, H.L. Wei, T.A. Palmer, A. De, et al., Building blocks for a Digital Twin of additive manufacturing, Acta Mater. 135 (2017) 390−399.

[18] R. Söderberg, K. Wärmefjord, J.S. Carlson, L. Lindkvist, Toward a Digital Twin for real-time geometry assurance in individualized production, CIRP Ann. Manuf. Technol. 66 (1) (2017) 137−140.

[19] T. Gabor, L. Belzner, M. Kiermeier, M. Beck, A. Neitz, 2016. A Simulation-based architecture for smart cyber-physical systems. IEEE International Conference on Autonomic Computing, 2016, pp. 374−379.

[20] T. Hey, S. Tansley, K. Tolle, The Fourth Paradigm: Data-Intensive Scientific Discovery, Microsoft Research, Redmond, 2009.

[21] J. Pirker, E. Loria, S. Safikhani, A. Künz, S. Rosmann, Immersive virtual reality for virtual and digital twins: a literature review to identify state of the art and perspectives. 2022

IEEE Conference on Virtual Reality and 3D User Interfaces Abstracts and Workshops (VRW), 2022, pp. 114–115.

[22] D. Frederica, DDDAS, a key driver for large-scale-big-data and large-scale-big-computing, Procedia Comput. Sci. 2015 (2015).

[23] F. Chinesta, E. Cueto, J. Duval, F. Khaldi, Virtual, digital and hybrid twin: a new paradigm in data-based engineering and engineered data, Archives of Computational Methods in Engineering, Springer Verlag, 2019.

[24] R. Khan, K.T. Ng, Model order reduction for finite difference modeling of cardiac propagation using DMD modes. 2018 International Applied Computational Electromagnetics Society Symposium (ACES), 2018, pp. 1–2. https://doi.org/10.23919/ROPACES.2018.8364207.

[25] S. Abburu, A.J. Berre, M. Jacoby, D. Roman, L. Stojanovic, N. Stojanovic, COGNITWIN — hybrid and cognitive digital twins for the process industry. 2020 IEEE International Conference on Engineering, Technology and Innovation (ICE/ITMC), 2020, pp. 1–8.

[26] A. Chambard, Deploying the Hybrid Twin to aid the automotive industry's digital transformation, ESI Group, October 18, 2021 (accessed 13.10.22). [Online]. Available from: https://www.esi-group.com/blog/deploying-the-hybrid-twin-to-aid-the-automotive-industrys-digital-transformation.

[27] R. Said, Hybrid Twin™ vs. Digital Twin: We'll Tell You the Difference and Which Can Save the Life of Your Asset, ESI Group, September 5, 2019 (accessed 13.10.19). [Online]. Available from: https://www.esi-group.com/blog/hybrid-twin-vs-digital-twin-well-tell-you-the-difference-and-which-can-save-the-life-of-your-asset.

[28] I.B.M. Fariz Saracevic, Cognitive Digital Twin. Technical Report, 2017. https://www.slideshare.net/BosniaAgile/cognitivedigital-.

[29] X. Zheng, J. Lu, D. Kiritsis, The Emergence of Cognitive Digital Twin: Vision, Challenges Twin.

[30] Ö. Albayrak, P. Ünal, Smart steel pipe production plant via cognitive digital twins: a case study on digitalization of spiral welded pipe machinery. Cybersecurity workshop by European Steel Technology Platform. Springer, Cham, 2020.

[31] P. Eirinakis, et al., Enhancing cognition for digital twins. 2020 IEEE International Conference on Engineering, Technology and Innovation (ICE/ITMC). IEEE, 2020.

CHAPTER SIX

Examples of energy, transportation, and information with humanity

Wei Han[1], C.C. Chan[2], George You Zhou[3], Zhiyong Yuan[4], Yingjie Tan[4], Hong Rao[4], Tik Lou[5], Jiawei Wu[5], Haohong Shi[5], Anjian Zhou[6], Changhong Du[6], Guocheng Lu[6], Yue Qiu[7], Suyang Zhou[7], Wei Zhang[8], Ying Li[9], Chunying Huang[9], Hailong Cheng[9], Mingxu Lei[9], Dan Tong[9] and Chi Li[10]

[1]Sustainable Energy and Environment Thrust, The Hong Kong University of Science and Technology (Guangzhou), Guangzhou, P.R. China
[2]The University of Hong Kong, Hong Kong SAR, P.R. China
[3]National Institute of Clean and Low-Carbon Energy, Beijing, P.R. China
[4]China Southern Power Grid Co., Ltd. (CSG), P.R. China
[5]XEV Company, Italy
[6]Deepal Automobile Technology Co., Ltd., P.R. China
[7]Southeastern University, Nanjing, P.R. China
[8]State Grid Integrated Energy Service Group Co., Ltd., P.R. China
[9]Institute of Digital Guangdong, Guangzhou, P.R. China
[10]International Academicians Science & Technology Innovation Center, P.R. China

Abstract

The development of a new type power system with new energy as the main body has become a major focus in the effort to create a more sustainable future. This new system requires innovative solutions that are not only environmentally friendly but also efficient and cost-effective. One such solution is the integration of behavior-driven smart vehicles and green transportation, which can significantly reduce carbon emissions and promote eco-friendly mobility. Smart electric vehicles are an important way to achieve low-carbon emissions and green transportation, with the drive system, Internet of Vehicles, autonomous driving, and smart cockpit being the four core areas of smart cars. The integration of co-robot-driven smart manufacturing plants and smart electric vehicles can lead to a more sustainable and efficient future. Collaborative robots can improve efficiency in manufacturing, while smart EVs can achieve low-carbon emissions and green transportation. User behavior plays a significant role in the success of smart EVs. The adoption of these solutions can reduce our dependence on non-renewable resources and promote eco-friendly mobility, leading to a more productive and self-sufficient society.

Intelligent integrated energy services (IES) is a new energy service system that takes electricity as the core, integrating gas, heating, transportation, and other energy systems. It relies on the continuous innovation of cogeneration equipment, energy conversion equipment, renewable energy, and other related technologies to achieve a high degree of synergy among multiple energy production, transmission, conversion, storage, consumption, and other aspects. With the help of IES, we can achieve a more efficient and reliable energy supply, reduce carbon emissions, and promote the development of renewable energy. The practice of four network four flows integration is an approach to digital Renaissance that can further enhance the efficiency and effectiveness of these innovative solutions. This approach involves the integration of information networks, logistics networks, capital flows, and talent flows, creating a seamless and interconnected system that can optimize resource allocation and promote innovation.

Overall, the combination of these innovative solutions can lead to a more sustainable and efficient future, with a reduced environmental impact and greater economic benefits. The practice of four network four flows integration can further accelerate this progress, paving the way for a digital Renaissance that can transform the way we live and work.

Keywords: New type power system; digital power grid; smart electric vehicle; green transportation; cobot; smart manufacturing; integrated energy system; intelligent integrated energy services; Digital Renaissance; smart city; four network four flows

6.1 Build the new type power system with new energy as the main body

The energy industry, especially the power sector, is the largest contributor to global greenhouse gas emissions. In 2019 carbon dioxide emissions from the energy sector accounted for almost half of the world's total emissions, with electricity and heating being the leading sectors. In 2021 emissions from these sectors increased significantly, highlighting the urgent need for a transition to a green and low-carbon development path.

To address this issue, China proposed the development of a new type power system in 2021, which aims to upgrade the traditional power system using digital technology, intelligent equipment, and the energy internet, among other things. This will facilitate the development of new energy resources such as wind and solar, and reduce reliance on traditional fossil fuels, ultimately achieving "carbon peaking and carbon neutrality" in the entire energy sector, particularly in the power industry chain. The proposal has gained widespread consensus among governments, utilities, research institutes, and academics in China, and detailed roadmaps have been formulated. For example, in 2021, China Southern Power Grid Co. Ltd. released a "White Paper on the Action Plan of Building a New Type Power System", and the State Grid Corporation of China also issued an

"Action Plan for Building a New Type Power System with New Energy as Main Body". These two main utilities in China have proposed plans to achieve their goals during 2030s.

6.1.1 Main features of the new type power system

With the development of a new type power system based on sustainable energy, significant changes are likely to occur in the functional positioning and morphological characteristics of power generation, transmission, distribution, and consumption. According to relevant research, the following transformations are highly probable:

1. **Power generation:** The establishment of a power supply portfolio based on sustainable energy will accelerate, resulting in significant changes in the positioning and role of various power sources. The installation capacity and power generation of new energy will experience sustained high growth, and their penetration level will increase significantly. The scenario of "wind and solar leads, multisource coordinates" may emerge, and new energy will gradually shift from a supplementary power source to a primary power source. Coal-fired power generation will lose its current position as the main power source and transform into a backup power source that mainly provides instant power support instead of long-term energy balancing. Hydro and nuclear power will maintain their fundamental roles as power and energy sources. Natural gas power generation will mainly play the role of regulating and securing power source. Pumped storage will continue to provide services such as peak-shaving and valley-filling and contingency backup.

2. **Power grid:** Extra-high/ultra-high voltage interarea power transmission and distributed power grids will be equally emphasized. The main structure of the grid will transform into a form of "reasonable partition, flexible interconnection, secure controllability, open complementarity," and the self-balancing capability of the provincial power grid will significantly improve. The power distribution network will be the basic unit for power balancing, with distributed power grids that have distributed generation and users as the primary body. Electricity will mainly be produced and consumed locally. The relationship between the transmission network and distribution network will change from master—slave to interdependence, and self-regulation can be achieved in a hierarchical partition manner. The power grid is expected to be highly integrated with other forms of energy networks,

such as gas, heating, and cooling, to form an "energy internet." The power grid will act as the energy hub for energy interaction and play a central role in energy transformation.

3. **Electricity consumption:** An integrated energy consumption system dominated by electricity will be established. User-level electricity/cooling/heating/gas/storage-based integrated energy systems will be widely implemented, continuously improving terminal energy consumption efficiency. The percentage of electricity in terminal energy consumption will grow dramatically, with electric vehicles and smart home appliances becoming more popular. A large number of loads will be transformed from "consumer" to "prosumer." Particularly, distributed generation, distributed energy storage, virtual power plant, and electric vehicles will be capable of exporting electricity into the power grid. The electricity market plays an essential role in balancing power supply quality and electricity cost. More users will participate in electricity market transactions directly or indirectly, balancing power supply quality and electricity cost through market measures.

These changes will bring significant challenges to power supply reliability, security, stability, and economic operation of power systems. Therefore systematic investigation and transformation are urgently needed in the areas of technology, economics, regulations, and policies. From the perspective of China Southern Power Grid Co. Ltd., a digital power grid is the solution for the new power system [1].

6.1.2 Empowering the new type power system with digital power grid

Digitization is a crucial trend for the transformation of power industry. Innovative applications of digital technologies are paving the way for building a new type power system. Driven by a new generation of digital technologies such as cloud computing, big data, internet of things, mobile internet, artificial intelligence, and blockchain, the digital power grid is a new energy ecosystem built on the base of a modern electric power network and a new generation of communication network, with data as the essential factor of production. By fully integrating digital technologies with the business and management of energy enterprises, the level of digitization, networking, and intelligence for power grids will be improved.

Power companies are actively utilizing technologies such as microsensors, edge computing, electric internet of things, and big data mining to build a powerful platform capable of cloud-edge coordination, massive data processing, data-driven analysis, intelligent decision-making, etc. The purpose is to identify operation patterns and potential risks by exploring data relationships on the foundation of massive data and intensive connections, rather than relying solely on power grid models. This approach ensures the secure and stable operation of the power system and the optimal allocation of resources. By constructing a digital power grid, power enterprises aim to provide power grids with sharper "sensory organs" and a smarter "brain," providing high perceiving, intelligent decision-making, and fast executing capabilities [2].

1. On the power generation side, the visibility, controllability, and accommodation capability of new energy generation can be substantially improved. With the use of multimodal data acquisition, functions and performance of traditional operation monitoring can be expanded, enabling power grids to have a comprehensive perception of new energy and meet the needs for real-time monitoring and precise forecasting of power generation from new energy. The application of technologies such as edge computing and Internet of Things (IoT) can perfect power grids' control capability over new energy, obtaining automatic generation control and fast frequency response, thus ensuring "friendly integration" of new energy. Additionally, the decision-making capability for dispatching new energy generation can be enhanced by technologies such as big data, cloud computing, and artificial intelligence, which support massive new energy generation to act as the main power source participating in power system regulation and control.

2. On the power grid side, transparency of the power grid can be significantly improved. Widely distributed sensing terminals and information communication networks enable system-level transparency of power grid status, device status, and management status. On this basis, through smart coordination and optimized dispatching, operation efficiency can be improved in power generation, transmission, distribution, and consumption sectors, contributing to comprehensive accommodation and efficient utilization of the new energy. By analyzing and diagnosing universal data, vulnerable points and operation risks can be detected, providing anticipated situation awareness and decision-making on operations, and ensuring secure operation of the power system with high penetration of new energy. Relying on transparent information, interaction can be enhanced between the power grid and upstream/downstream participants of the power industrial chain as well as electricity consumers, forming a more efficient, greener, and more economical energy ecology.

3. On the electricity consumption side, a modern service system can be established that accommodates the requirements of a new power system. With the use of technologies such as the IoT and blockchain, massive dispatchable resources on the user side can be aggregated, making virtual power plants possible and promoting two-way interaction between generation and consumption. By applying a flexible digital electricity exchange platform, users' requirements for clean and convenient power supply services can be met. In addition, value-added services such as electricity transactions and energy efficiency management can be enabled, contributing to the development of new energy services such as distributed generation, electric vehicles, electricity substitution, and supply–demand interaction.

In summary, by constructing a digital power grid, power enterprises can improve the level of digitization, networking, and intelligence for power grids, ensure secure and stable operation of the power system, and contribute to the comprehensive accommodation and efficient utilization of new energy. With the application of digital technologies, power grids can be transparent, efficient, and interactive, forming a more efficient, greener, and more economical energy ecology.

6.1.3 Intelligent integrated energy services

The integrated energy system (IES) is a new energy service system that integrates electricity, gas, heating, transportation, and other energy systems [3]. It relies on the continuous innovation of cogeneration equipment, energy conversion equipment, renewable energy, and other related technologies to achieve a high degree of synergy among multiple energy production, transmission, conversion, storage, consumption, and other aspects. In the context of the energy Internet, the development of measurement, communication, Internet, and control technologies provides essential technical support for information interaction and cooperative control among heterogeneous energy flow systems such as electricity, natural gas, and heating. The diversified and intelligent energy services provided by IES to serve people's needs for a better life and efficient productivity are called intelligent integrated energy services (IIES). The IIES encompasses various energy businesses, including energy infrastructure construction, sales, distributed energy, energy efficiency, demand response, and financial derivatives. IIES reduces society's dependence on traditional fossil fuels while promoting renewable energy consumption. It facilitates the integration of "four networks and four flows," breaking down barriers between heterogeneous energy networks such as electricity, gas, and heat. IIES offers alternative energy supply and consumption options, like switching between natural gas stoves and electric cookers, multiheat source complementation between gas boilers and heat pumps, and transportation electrification. Additionally, clean energy carriers like hydrogen are emerging, replacing fossil fuels through electric hydrogen production, fuel cells, and hydrogen–enriched compressed natural gas (HCNG) technologies. Recent advancements in building energy conservation and intelligent energy management, such as near–zero energy buildings and smart towns, improve demand-side self-sufficiency. IIES integrates humanities and energy networks, providing diverse energy usage, transportation, and cultural lifestyles. It empowers users to become prosumers, and new energy service providers emerge, such as energy storage service providers and load aggregators. Humans can adjust their energy use behavior and methods in response to incentive and punishment signals, positively interacting with the energy network to serve its needs. IIES has three fundamental features: multienergy flow, multiagent, and multiscale, reflecting its rich connotation from different angles. The "multienergy flow" feature distinguishes IIES from traditional independent power, heating, and natural gas systems. Meanwhile, IIES's electricity, gas, and heating subsystems have their own flexibility resources and demands, but their physical characteristics may differ. The power system focuses on long-term energy and short-term power balance, while the heating system considers long-term thermal energy supply and demand balance, as well as short-term heating economy and environmental protection. Synergy and complementarity among multiple energy flows can be seen as virtual energy services, where flexible resources are fully utilized to provide energy services for diverse users.

"Multiagent" means that the flexible resources mobilized by IIES involve different stakeholders. On the one hand, electricity, gas, and heating subsystems usually belong to different departments. Typical core demands of various departments include economic efficiency, green and low-carbon energy supply, and energy supply safety and adequacy. The improvement of certain performances of one entity may be at the expense of other entities. Therefore IIES between different entities is accompanied by the transfer of costs. On the other hand, the flexible resources of each link may belong to independent service providers, such as energy storage service providers and load aggregators. Then, flexible resources become a unique commodity in the energy system, and the realization of its value depends on an effective market mechanism. To sum up, the "multiagent" feature means that some flexible resources in the IIES are difficult to allocate in a unified manner, and a suitable market mechanism is needed to explore their potential.

In addition, the "multiagent" feature also brings problems such as information barriers. "Multiscale" refers to the multitime-scale and multispatial-scale features of the IIES. On the time scale, energy demand has different performances in a year, quarter, month, day, and real-time. The focus is gradually shifted from the mid- and long-term regulation capability of system energy to the short-term or real-time regulation capability of system power. On the spatial scale, according to different application requirements, IIES can not only focus on various building users (such as energy-saving and fee-reducing services of the smart home) but also can be expressed as city-level and even regional-level service capabilities (such as regional demand response provided by aggregators for external grids).

6.1.4 The basic theory and methods for intelligent integrated energy services

The core technical support of IIES can be summarized into three aspects: modeling and simulation, planning and design, and operation optimization. Modeling and simulation are the basis of the operation, planning, and analysis of IES. IES modeling can be further divided into equipment modeling and network modeling. Equipment modeling includes models of generators, various prime movers and energy storage systems in power systems, models of gas transmission stations, pressure regulating stations, and gas storage systems in natural gas systems, as well as models of heat sources, heat exchangers/heat exchange stations in heating systems. Network modeling involves power distribution and transmission networks, gas distribution and transmission networks, and heating supply networks. The IES simulation can be divided into three parts: static energy flow calculation, dynamic energy flow calculation, and dynamic simulation analysis considering the dynamic characteristics of the equipment. The purpose of the static energy flow calculation is to obtain the system's operating parameters at a certain time section according to the load demand and the output of the source node to provide a basis for the safe and economical operation of the energy system. Based on static energy flow, dynamic energy flow calculation further considers the unique delay characteristics of energy networks such as heat and gas grids in the energy transmission process and obtains the changing laws of parameters such as power and flow over time. Based on the dynamic energy flow calculation, the dynamic characteristics of the system components are further considered, which constitutes the dynamic simulation analysis of IES.

Unlike the traditional energy system's planning of single energy forms such as electricity, gas, and heating, the planning and design of IES need to break down industry barriers, remove policy and geographical restrictions, and give full play to the complementary advantages of energy sources. Scientific and reasonable IES planning can meet the needs of the planning target in various aspects, such as economic benefits, renewable energy consumption, energy saving, and emission reduction. IES planning needs to consider the resource endowment, economy, policy, society, environment, and other factors of the planning area. The designed equipment-level and network-level power, gas, and heating subsystems must meet the planning area's multienergy load demand. Stochastic programming, robust programming, multiobjective programming, and multistage programming are the commonly used IES planning methods.

IES operation optimization aims to determine the optimal output of the equipment in the system and the optimal power distribution of the network to optimize the system operation economy, environmental protection, or energy efficiency. For a given load profile, the system often has a variety of feasible operation schemes, but different schemes may have significant differences in performance. Therefore operation optimization is the key to exerting the efficiency of IES. From a physical point of view, the energy flow of electricity, gas, heating, and other energy flows in an IES presents a complex coupling relationship through energy conversion equipment. The redundant configuration of the system and the complementary characteristics of various energy flows bring relatively high operational flexibility. In contrast, factors such as renewable energy output, multienergy load, and meteorological factors that are difficult to predict make the system operation face multiple uncertainties. To summarize, IES operation control is a complex problem with multienergy flow coupling, multitime scales, and multioperating objectives.

6.1.5 Typical intelligent integrated energy services platforms

In terms of the IIES platform related to modeling and simulation, the separate power, gas, and heating simulation modules have developed rapidly. The mature commercial software includes the TRNSYS, Modelica, and TERMIS, which can perform heat network simulation; the DIgSILENT/PowerFactory, pandapower, PSSE, and DGRSS, which can perform power simulation; and the Pipeline Studio, which can simulate gas systems. However, commercial software to support integrated energy cosimulation has not been effectively developed. The TRNSYS only supports the simulation of a limited number of devices and networks in electrical, cooling, and heating networks. In 2020 Southeast University developed the IES-Sim, a simulation platform for electricity–gas–cooling–heating–steam IES equipped with more than 50 devices and 5 network models, but it has not yet been commercialized.

In terms of IIES platforms related to planning and design, several platforms, including the HOMER, DER-CAM, CloudPSS-IESLab, and IES-Plan, have been developed at home and abroad in recent years. The DER-CAM and HOMER were developed by Lawrence Berkeley National Laboratory and National Renewable Energy Laboratory (NREL), respectively. The CloudPSS-IESLab and IES-Plan were developed by Tsinghua University and Southeast University, respectively. Among them, the DER-CAM and HOMER focus more on the planning and evaluation of power systems, while the CloudPSS-IESLab and IES-Plan consider the coupled planning and simulation of multiple energy forms and multiple integrated energy equipment. Regarding the IIES platform related to operation optimization, the relatively mature IES energy management platform at home and abroad is the IEMS developed by Tsinghua University in 2020.

6.2 Behavior-driven smart vehicle and green transportation

Green transportation is not only a strategic layout made by countries around the world in the face of the fossil energy crisis, but also a travel mode chosen by modern consumers based on the concept of environmental protection. This is the combined result of macro-control and ordinary consumer behavior. Smart electric vehicles (EVs) are an important way to achieve low-carbon emissions and green transportation. Drive system, Internet of Vehicles, autonomous driving, and smart cockpit are the four core areas of smart cars, driven by user behavior. This section analyzes the impact of user behavior on the development of smart EVs. When smart EVs can solve some of the limitations of traditional EVs and shared transportation, more users can be attracted to use smart EVs, and even abandon traditional fuel vehicles to achieve green transportation.

6.2.1 Introduction

Green transportation refers to a travel mode that causes less environmental pollution. The future transportation mode will be mainly defined by users, that is, user driven. We predict that the main means of transportation in future cities will be zero-emission smart electric vehicles (EVs), and a public transportation system using EVs, because vehicles will greatly contribute to the realization of low-carbon emission green transportation. To achieve green transportation, promoting the development of smart EVs will play an important role. Only when car-sharing and other public transportation services are highly intelligent and can provide users with a high-quality travel service experience, users will be more inclined to use public transportation, thereby reducing the number of traditional fuel vehicles. Smart cars are intelligent upgrades based on electric cars. The development of electric cars is the premise and foundation for the development of smart cars. At the same time, the development of smart cars will further promote the popularization of electric cars and change users' inertial dependence on cars with traditional internal combustion engines. In the development of smart EVs, electrification and intelligence are two important aspects.

Electrification refers to the extensive application of power electronics technology in automobiles, and the drive mode is changed from traditional internal combustion engine drive to electric motor drive. The electric drive system is an important symbol of vehicle electrification. The electric drive system mainly includes motor controllers, motors, reducers, power modules, and other components. In the past 10 years of development, the drive system at home and abroad has shown a trend of high integration. More and more components and subsystems are integrated into the drive system. These power drive systems can be collectively referred to as an all-in-one drive system. At the same time, the development trend in the chassis field is to replace the independent controller electrical park brake (EPB) brake system with integrated electric park brake (EPBi) electronic brakes. It is estimated that the scale of the domestic market for brake-by-wire, steering and suspension alone will exceed 80 billion yuan in 2026. The average compound growth rate is 45%−62%. More and more mid-to-high-end products have started to use air springs and semiactive suspensions, and new motors, electronic valves, or control pumps will be introduced as actuators. Finally, the number of electrical equipment on board has also increased, such as on-board radar, electronic camera, nonsensing entry system, augmented reality head-up display (AR-HUD), and other electronic equipment. The increase in the type and quantity of electrical equipment and the increase in control chips are notable technical features of the electrification revolution.

Intelligence is to add software and algorithms based on electrified hardware to build an on-board artificial intelligence. The form of smart vehicles not only depends on the manufacturer's research and development, but also is affected by user behavior, usually termed behavior driven. Autonomous driving, smart cockpits, and the Internet of Vehicles (IoV) are the three main areas of development for smart cars. The development of autonomous driving is divided into five levels, starting from L1 to L5, where L5 represents the highest level of intelligence. At present, affected by the level of IoV and the imperfect vehicle to everything (V2X) technology infrastructure, the main original equipment manufacturer (OEM) equipment is at the level of L2-L3, leaving a large space for development. Autonomous driving focuses on the driving experience, reduces the requirements on the driver, and allows the driver to free his hands during driving. The smart cockpit focuses on the riding experience, improves the safety and entertainment of the vehicle, provides the driver with vehicle information display, control, and seat settings, and adds a new driving experience mode. IoV is the basis for the information collected by smart vehicles, and it is also a key service that autonomous driving and smart cockpits rely on. This chapter classifies the development level of smart cockpits according to the interaction with users and predicts the form of smart cockpits in the future.

Whether it is electrification transformation or intelligent design, the logic of product design development remains unchanged. It starts with user behavior, analyzes user needs, plans product functions and features, completes product technical detailed design, and sets product performance indicators. The tasks in the integration stage of parts and subsystems, include checking and accepting product performance, carrying out test verification at all levels of the product, testing and verifying the function and performance of the subsystem, conducting real vehicle road testing, and finally analyzing the user feedback after the product is launched into the market. The above V-shaped development logic is widely used in industry. User needs are defined by user behavior, and user behavior is correct, wrong, standardized, irregular, and even micro-individual and macro. User behavior determines user needs and has always affected the functional changes and technical features of the final product in terms of electrification and intelligence. In short, user behavior will determine the form of smart cars and green travel.

6.2.2 Future trends of electric vehicles

6.2.2.1 Electrification

Part of the trend in vehicle electrification is the use of batteries and/or fuel cells as power sources, and another part of the trend in electrification is the application of electronic components (such as sensors and microprocessors) in vehicles, such as parking radar and imaging, cabin lighting and audio systems, electronic chassis, and electronically controlled suspension. The electrification trend is partly due to the higher energy efficiency of electric motors (85%—95%) compared with internal combustion engines (30%—45%) [4,5], higher energy efficiency means that the same mileage consumes less energy. After electrification, the use of electric energy is no longer dependent on fossil fuels, and there are a wide range of sources of electric energy, including hydropower, solar power, nuclear power, and wind power. The proportion of clean energy will gradually increase, reducing the use of thermal power generation, which in turn will reduce harm to the environment, and reduce the excessive dependence on oil resources. At the same time, the electric motor has no exhaust gas emission, which is of great significance for improving the air quality in urban life. As a representative of vehicle electrification, EVs will gradually replace internal combustion engine vehicles to reduce carbon dioxide emissions.

6.2.2.1.1 The electric vehicle drive system

The main trend of the electrification of the drive system is high integration and high performance, with high power density, high voltage, high efficiency, high frequency, low noise as the main features. After the battery pack replaces the fuel tank, the vehicle has the ability to use higher power electronic products, making electrification transformation possible, so it can provide scene services and functional designs that are impossible or difficult to achieve in the era of fuel vehicles, and introduce many new electronic appliances, such as AR–HUD, various radars, and cameras. Users' pursuit of performance and intelligence will lead to product performance requirements and new service experience requirements, which is the underlying logic of user-driven design. The highly integrated drive system can use an integrated housing to reduce the installation of parts, and reduce the mass and volume while ensuring the durability of the structure through lightweight technology, thereby increasing the power density. The drive system has more additional functions, adopts higher power supply voltage and switching frequency, and the new power architecture has higher overall efficiency and lower overall noise level.

From three-in-one to seven-in-one, and then to multi-in-one drive system, has proven to be a successful development path. High integration requires overcoming multiple technical difficulties, including reliability, thermal damage, vibration and noise, electromagnetic compatibility, efficiency, etc., all under the condition that the cost will not soar. Wide use of surface mounted devices (SMD) components helps to further reduce the scale of control and signal acquisition circuits, thereby achieving high integration. High voltage refers to the use of higher voltage on the direct current (DC) side of the inverter. At present, 240—480 V is common. Upgrading from 240 to 800 V is a technically feasible development route. At present, many electric drive system products that withstand 800 V or more have been launched on the market. The use of 800 V high voltage will increase the charging power, greatly reduce the charging and recharging time, and bring convenience to people's travel. The motor adopts a three-phase power supply. The higher voltage level usually means a wider constant torque area. If the rated power of the motor is made larger, the equivalent constant power area will be wider, which is very important for improving the power performance of the vehicle. In the process of driving the vehicle forward, the total efficiency of the drive system can be improved by designing a new cooling structure, identifying the core parameters of the system at different temperatures online, adjusting the switching frequency of power electronic components, and using the optimal efficiency control scheme. The number of connecting harnesses and components is reduced, improving system reliability. The frequency range that the human ear can perceive is 20—20 kHz. The higher the noise frequency, the easier it is for the sound to be absorbed and cut off. Currently, commercially available switching components are mainly Si-based insulated-gate bipolar transistors (IGBTs) with a switching frequency of 5—20 kHz. Gallium nitride (GaN), under research in academic circles, can be used as a substitute for Si-based materials, with a greatly improved switching frequency, enabling 650 V/100 MHz components. The noise characteristics of the motor can be further improved if the higher switching frequency is used and the motor is controlled by space vector pulse-width modulation (SV-PWM) technology. At present, pure EVs widely use electronic gear shifting, using keyboard keys, knobs, and levers to manually switch to power gear, reverse gear, and neutral gear. This is also a change in control after the electrification transformation. These technical developments and performance characteristics are all new challenges and new situations brought about by the electrification revolution (Fig. 6−1).

User behavior is the driving force behind product design. During the electrification transformation process of the drive system, the industry generally borrowed the "V-shaped" design and verification concept, forming a closed-loop development logic for product design and product performance. This driving logic is always customer-centric, thinks from the perspective of consumers, converts service requirements into functional requirements, and then decomposes functional requirements into systems, subsystems, and components, and also forms technical requirements for components. Component technical requirements, subsystem, and system-level technical requirements, after product integration, meet the technical requirements in the harsh verification environment of "alpine cold, high temperature, and high humidity," and qualify the drive system under the "seven horizontal and seven vertical" verification schemes, functional design, and comprehensive performance goals. Before the R&D products are launched to the market, at least three rounds of internal closed-loop verification design will be formed in the internal R&D test verification system. After the products are introduced to the market, users' actual usage opinions will be further fed back to the technical department to form an external closed-loop iterative design and realize the product iteration logic driven by users (Fig. 6−2).

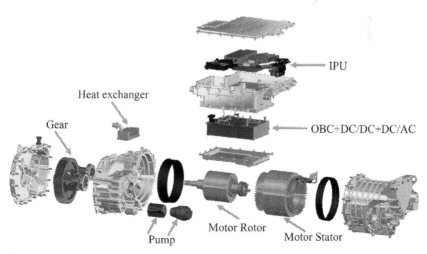

Figure 6−1 All-in-one electric drive product structure explosion diagram.

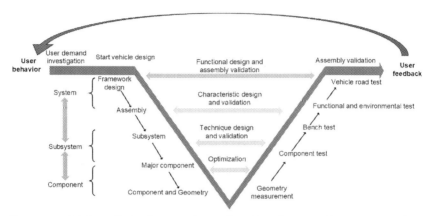

Figure 6–2 "V-shaped" development logic of product design in the automotive field.

6.2.2.2 Intelligentization

Nowadays, people are no longer satisfied with just using a car as a pure driving tool. Software, chips, and other electronic devices endow vehicles with unprecedented intelligence and capabilities, especially the intelligent design of the vehicle end can effectively empower the vehicle and increase the intelligent experience mode. At present, in a normal passenger car, there are about 500–600 pieces of various microcontroller units (MCUs) or chips in traditional fuel vehicles, while the number in pure electric cars is about 1500–2000, which is 1.5 times more. With the continuous application of chips, the number of chips may be further increased, which represents the level of electrification and intelligence. In the future, we may no longer need to manually operate the vehicle in the cockpit, but the vehicle will transport us to our destination autonomously and safely, and we can enjoy concentrated and comfortable online entertainment or office work. When we are enjoying the fun of driving, smart vehicles can automatically correct wrong operations and protect our safety. After the smart car is connected to the Internet, especially after the IoT is more developed, the information communication between objects can be strengthened. The vehicle can obtain the basic status and information of the surrounding moving objects in real time, intelligently identify the optimal driving route, and avoid congestion. Due to the waiting delay caused by too many traffic lights, the optimal power supply scheme is obtained through calculation and analysis, which keeps the battery in a good charge and discharge state and prolongs its service life. Some of the current household electrical equipment has realized automatic power replenishment and intelligent charging functions. In the near future, EVs will be able to go to the charging station for wireless charging. After hydrogen energy fully enters society, the hydrogen refueling station may not be as good as the current one. With the popularization of charging stations, intelligent and automatic hydrogenation of vehicles will also become a reality during the transitional period, thereby reducing human intervention; in the case of insufficient hydrogenation stations, all kinds of vehicles after networking will be recharged after obtaining user authorization. In some cases, queuing hydrogenation can be carried out. Human—vehicle interaction will increase, new interactive contacts and modes will be extended, vehicle-to-vehicle interactions will increase, vehicle-to-vehicle distance and optimal driving planning routes will be maintained, and the interaction between vehicles and urban intelligent transportation systems will also increase. In the traffic indication system represented by traffic lights, people will travel intelligently in a standardized, orderly, and efficient manner.

The intelligence of the drive system is also a development trend. The intelligent transformation design of the drive system is reflected in two aspects:

On the one hand, the drive system carries the will and mission of the intelligent engine, and is the terminal executive mechanism, with intelligent tentacles. As the hands and feet of the intelligent brain, the drive system will execute instructions accurately and quickly, and realize a series of control strategies such as drive, energy recovery, heating, charging, and entertainment functions. These smart strategies are like the brains of the drive system, and ultimately control the vehicle to achieve people's wishes.

On the other hand, the intelligent response of the control system is reflected in the abnormal working mode and failure mode. From these two perspectives, the drive system is not only an energy conversion device, but also a management center with certain "intelligence." By monitoring the operating status of the drive system, real-time analysis of data streams such as temperature, torque, current, voltage, motor speed, and vehicle speed, the functional safety diagnosis can be completed online to avoid abnormal operation of the drive system and its subsidiary subsystems, or to prevent the drive system from causing damage to the battery system. Potential hazards of overcharging and overdischarging or avoiding potential handling and stability hazards caused by the loss of grip of the chassis control system. In the configuration of high-end passenger cars, the joint use of intelligent driving technology can easily realize adaptive cruise and avoid driving when fatigued; active braking function can avoid rear-end collision; under the intelligent driving ability of L4 level, it has active obstacle avoidance function to avoid collision harm. But for users, the perceived changes are few and may not be obvious, and they can only perceive the intelligent results of some application scenarios on the car terminal display. The transformation at the level of cloud intelligence, vehicle-side application, interface, bottom, and physical layers cannot be perceived. Because these changes are driven by high reliability, high security, and high redundancy systems. The drive system design logic is under normal operating conditions, and no stress response will be triggered. Under normal and abnormal working conditions, the system can continue to output torque as much as possible and control stably to ensure system and personal and property safety. In exceptionally abnormal working conditions, technical measures have been formulated to maximize the protection of personal and property safety. These technical measures use advanced sensor technology, combined with high-speed bus technology, high-performance MCU, use centralized or distributed domains to control various subsystems, implement intelligent control and safety protection strategies, and improve the drive system's ability to perceive the entire vehicle. The cooperative control capability of each subsystem reflects a high level of intelligence. At present, some problems have been found in the development process of domestic independent products, especially with the stimulation of cloud data and the demand for intelligent transformation. The bus technology used in the traditional distributed domain control system architecture has fatal shortcomings, the amount of transmitted data is too small, the transmission and the data rate are too low to meet the requirements of cloud intelligent control and intelligent driving. Research shows that with the demand for intelligent development and OEM cost reduction control, full-stack development based on E/E architecture will become a development trend, which will affect the architecture design and manufacturing of intelligent development. It will put forward a new test for the OEM's cloud intelligent platform technology.

6.2.3 The development of smart vehicles

6.2.3.1 Internet of Vehicles

The concept of the IoVs is originated from the Internet of everything. This interdisciplinary concept contains multiple fields of technologies such as automobile, communication, and electronic [6]. Generally, the IoVs will be composed of an intra-vehicle, a vehicle-road, and an on-board mobile network. The intra-vehicle network links various sensors and electronic control unit processors in the vehicle to collect the working status of the vehicle and basic surrounding environment information to ensure the stable operation of the vehicle; the vehicle-road network is the information exchange between the vehicle and other units on the road, including other vehicles, road facilities, bicycles, and pedestrians; the on-board mobile network allows a vehicle as a terminal of the Internet for online data exchange. The IoVs is a necessary condition for in-vehicle artificial intelligence to obtain information, and it is also a key service that autonomous driving and smart cockpits rely on. The IoVs can transfer the traffic information network composed of people, vehicles, and roads during driving to artificial intelligence for planning, achieve safe and efficient travel, and save unnecessary energy consumption.

The architecture of the IoVs system includes environment sensing, network access, coordinative computing, and application. The environment sensing layer is hardware-based information collection. On-board and on-road data are acquired by sensors, radars, and IoT devices such as roadside units (RSU), radio-frequency identification (RFID), near field communication (NFC), and other personnel IoT devices. The network access layer allows uploading and sharing the information which collected by the IoVs terminals in a sensing layer through the vehicle mobile network, including dedicated short-range communication (DSRC) and Bluetooth for the vehicle-road network; wireless local area network (WLAN) and fourth/fifth generation mobile networks (4G/5G) for on-board mobile network; controller area network (CAN) and local interconnect network (LIN) for intra-vehicle network. In the coordinative computing layer, cloud computing service based on data center and vehicle computing unit (VCU) provides computing and analyzed results of the uploaded data for the human−vehicle−environment coordinated system. The application layer is the services for user layer based on the IoVs technology, and directly shows the functions and forms of IoVs, such as on-board entertainment, autonomous driving, safety, and other customized functions. IoVs services are designed and arranged by the provider according to the needs and expectations of customers. Since the services provided in the application layer are determined by the market and consumer preference, the development of the IoVs is driven by users' behavior. At present, the IoVs is still in the conceptual stage, information security and hardship of cross-industry cooperation are the challenges,

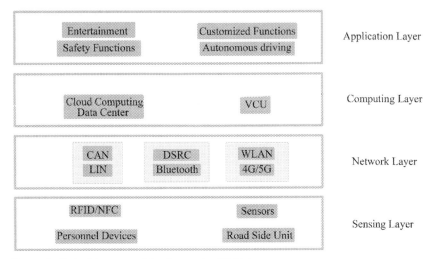

Figure 6–3 Architecture of the Internet of Vehicles.

restricting IoVs development. The software and hardware related to the IoVs exist in diverse types, sources, and specifications. Different standards cause compatibility problems and a high potential of hacker attacks. Besides, the expected large amount of IoVs terminals will provide massive data. The difficulty of designing IoVs software and algorithms is how to fast process data and offer real-time service for the user. In addition, the private information in the collected data also lacks standardized supervision. Enterprises from various part of the industry chain have different interests, IoVs industry should be looking forward to a business mode that can satisfy most of the companies at the same time. The strong demand from enterprises and consumers for IoVs will promote a rapid and standardized development (Fig. 6–3).

6.2.3.2 Autonomous driving

Autonomous driving technology in smart vehicles is achieved by the perception of the surrounding environment and road condition information through on-board sensors and IoVs devices, and then the artificial intelligence decision-making control system manipulates driving activities including steering and speed changes of vehicles to complete driving tasks on the road automatically [7]. As a revolutionary innovation in the improvement of driving experience, autonomous driving can partially or even totally replace human drivers as well as making driving safer. According to the National Highway Transportation Safety Administration of United States, around 94% of traffic fatalities related to driving in the United States are caused by human drivers' mistakes. Autonomous driving can reduce the mistakes in emergency situations when the human drivers might make wrong decisions under pressure. In addition, human drivers will no longer be required to concentrate on manipulating the vehicle with the help of autonomous driving, they can sit leisurely, shop, work, and be entertained inside the vehicle cockpit. In addition, some elderly people, minors, and passengers who do not have the ability to drive will have a chance to travel by vehicle independently.

According to the definition of the Society of Automotive Engineers (SAE) of United States, the development of autonomous driving can be divided into six levels. Level 2 autonomous driving has been applied to different models of multiple automakers as a proven technology. As the synonymous with new technology, the proportion of EVs equipped with autonomous driving functions is higher than that of fossil fuel vehicles. According to market research from the International Data Corporation in 2022, the market share of Level 2 autonomous driving of EVs is 35.0%, while the fossil fuel vehicles is 19.9%. Consumers of all ages have a certain understanding and acceptance of autonomous driving, especially for young EV users (Fig. 6−4).

Autonomous driving in vehicles will be developed from full control by a human driver, to conditional driving automation, and then will finally achieve full automation. The development process of autonomous

Figure 6−4 Levels of driving automation, represented from Society of Automotive Engineers Standard J3016 (Taxonomy, S. A. E. "Definitions for terms related to driving automation systems for on-road motor vehicles. Publication J3016_202104." Society of Automotive Engineers (2021)).

driving requires the willingness of users to try the system. Only when the user is continuously learning and adapting to autonomous driving, will the developer be able to continually collect feedback from users and improve the technology. However, the current autonomous driving technology is still in the early stages of development, consumers are skeptical about the reliability of autonomous driving technology and unwilling to hand over some control of their vehicle. In the research and development process, the autonomous driving system should be designed as user-led so that the drivers can trust this life-safety-related technology and hand over control of the car [8]. The impact of riding and operating a car which equipped with autonomous driving technology on the driver's psychological expectations and operating behavior needs to be evaluated during the research and development process. Different from the focused operation of the driver of a traditional car, the drivers of an autonomous vehicle require to take over control of the vehicle at any time from monitoring and possible nondriving activities such as entertainment under the current relatively low level of autonomous driving technology. This situation usually happens when an accident is about to occur and requiring the driver of the autonomous vehicle to make decisions and respond within a short period of time. If the driver of the autonomous driving vehicle overtrusts and overrelies on this technology under the early stage of the development, they might not concentrate on driving and not be able to rapidly take over the car in an emergency, which will eventually lead to accidents [9]. The accidents happened on autonomous driving vehicle will cause the mistrust and even opposition of users on autonomous driving, hindering the development of the technology. The visualization display system for users associated with autonomous driving detects objects and their predicted motions and can help users to understand the action of the autonomous driving system and enhance trust. Besides, the establishment of trust in the autonomous system in humans is also important when level of autonomous driving is low, in order to ensure that the control of the vehicle can be returned to the driver as soon as possible. When the system detects that the driver's state does not meet the conditions to take over the control of the car, the automatic driving system can re-awaken the driver's attention through sound and image prompts, or park the car in a safe place. In summary, the development of autonomous driving is driven by the behavior of users. The developers of autonomous driving technology should continue to pay attention to the impact of any function on the user and continuously upgrade the technology through the feedback of users.

Such developed route can build the two-way trust between humans and autonomous driving system, and finally reach full automation.

6.2.3.3 Smart cockpit

The cockpit in vehicle is the place where the user directly interacts with the vehicle. Under the trend of digitalization, intelligence, and networking, the human—computer interaction interface and human—vehicle interaction mode of automobiles have evolved into smart cockpits. The ways of human—vehicles communication and presentation of information are diversifying. As the carrier of software and hardware interaction, the physical components such as central control, instrument panel, and steering wheel in the smart cockpit are gradually being replaced by various digital interfaces. The updates of driving information display system and on-board entertainment system can improve the user's ride and driving experience, providing more possibilities for the diversity of automotive digital experience [10,11].

Since the current autonomous driving technology is not yet perfect, the vehicle still must be controlled by the human drivers. The smart cockpit is required to coordinate with autonomous driving technology in this period to improve safety. The safety functions of the smart cockpit include sensing of the surrounding environment outside the vehicle based on IoVs service and multisensory collaborative human—vehicle interaction assistance. The surrounding environment contains road conditions, information about pedestrians and other vehicles. At present, the collection of the information is challenged by real-time monitoring and computing by an efficient algorithm, a breakthrough after the further development of the IoV is expected. Human—vehicle interaction assistance can monitor the status of drivers and passengers, present the information by multiple senses (vision, sound), and allow users to interact with vehicles with new manipulation methods such as faces, gestures, eye tracking, and voice. After the safety requirements of users is reached, smart cockpits also need to load entertainment functions. Since the attention of the drivers should be concentrated, right now the entertainment functions are mainly focused on rear-seat passengers. After the development of autonomous driving to high level, the immersive entertainment system in the cockpit can be built, and with the on-board 5G mobile network, the vehicle can be seen as a mobile living space. In addition, diversified user demands will bring more diversities of form to the smart cockpit. Smart vehicles will become a data-integrated, passenger-centric mobile smart carrier.

The design of smart cockpit could be focused on the users' preference and interest rather than functions. More customized, more emotional, and more humanized in-vehicle interaction design will improve the user experience and satisfaction.

The keyword of the smart cockpit is human—vehicle interaction. In this chapter, we classify the development stages of smart cockpits which refers to the developed levels of artificial intelligence and autonomous driving (Fig. 6—5).

L1 passive interaction: The electronic equipment in the cockpit is an extension of the scene of consumer electronics products and passively waits to interact with the user, an example is voice assistants that can only recognize specific words. Users need to be trained and adapted to this level of human—vehicle interaction methods.

L2 semiactive interaction: The application of advanced equipment and algorithms such as eye tracking and language meaning recognition enhances the data collection and interaction capabilities of the smart cockpit. The supporting artificial intelligence algorithm allows the smart cockpit to remember the user's needs after the first operation, and actively convey information through vision and hearing in such scenes to realize the interaction with the user. Interaction modes are simplified; users do not need to be trained but should still follow specific interaction logic.

L3 active interaction: More and more complex sensors and processors are applied in the smart cockpit. The smart cockpit can autonomously

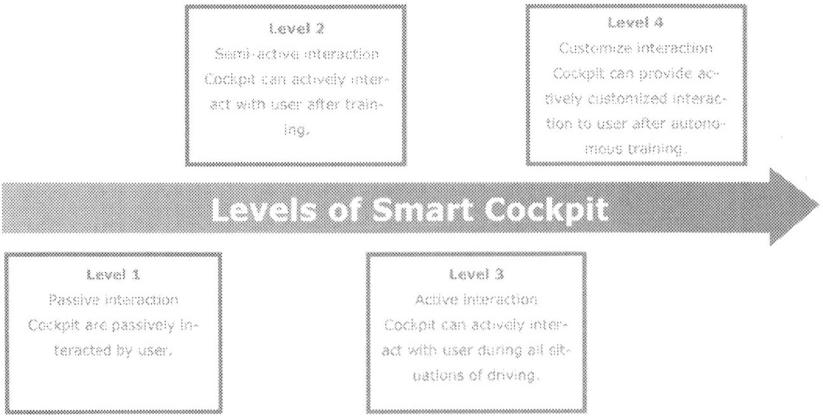

Figure 6—5 Levels of smart cockpit in smart vehicles.

process all information related to the potential requirements of users inside the cockpit and around the vehicle. The smart cockpit can independently provide interactions and services in all driving and riding conditions. Users do not need to be trained but should still follow specific interaction logics.

L4 customized interaction: The smart cockpit can provide customized interaction modes and services for users by learning the user's behavior and preferences. Through the cooperation of the IoVs terminals and other smart devices, the scene of living space-smart vehicle-working space can be seamlessly switched without barriers. Therefore the smart vehicle can be turned into a third living space.

The development of smart cockpits is also behavior driven. The hardware and software, and the core functions of the smart cockpit, need to be upgraded continuously during interactive training with the user for better matching of the user's preferences and the formation of an operation style which is exclusive to the user. At the same time, users will also better adapt to smart vehicles as an extension of living space and use the functions of smart cockpits better. In the development process of the smart cockpit, the function and interaction design also need to focus on the impact of user behavior by studying the human—computer interaction of smart vehicles. When designing an in-vehicle entertainment system in a smart cockpit, cognitive load, driving safety, and human—vehicle interaction cognition are the keys to balancing driving behavior and cockpit human—computer interaction. From the perspective of software and hardware interaction, the infotainment system makes the user's interaction task flow more complicated and puts forward higher requirements for the drivers. This might cause the overload of human—vehicle interaction cognitive load during driving and affect the driver's ability to cope with emergencies and situational awareness and handle multiple interactive tasks, and the ability to deal with a variety of complex situations [12,13]. From the perspective of user experience during the process of driving, the human—computer interaction tasks inside the cockpit will lead to the distraction of the user's attention because the user needs to focus on the cognition of the environment, so that the user is prone to misuse in the distracted state and thereby reducing the human—computer interaction user experience during driving. Complex human—computer interaction tasks will inevitably affect user cognition, thus directly affecting the experience of using smart vehicles. Human—vehicle interaction needs to make a balance between driving tasks, human—computer interaction tasks in the cockpit, and user needs. Therefore the establishment of a smart cockpit interaction system needs to meet the basic cognitive habits and cognitive rules of drivers, so as to meet driving safety and provide a variety of user experiences when the current level of automatic driving technology is low. At present, there are more and more visual task-based interactive tasks for in-vehicle functions and infotainment systems, which put forward higher requirements for the driver's learning ability and the usability design of the human—machine interface. Interaction habits are turning to interfaces and voice-based interactions. The information presented in the smart cockpit in vehicle can be presented in a variety of ways, without being constrained by the physical performance of traditional physical buttons and display screens. Developers of smart cockpits need to cater to the diverse user needs of consumers in a more natural way of human—computer interaction, and deal with and satisfy the goals and motivations of users.

6.2.4 Conclusion

The three main areas of smart vehicle development are the IoVs, autonomous driving, and the smart cockpit. The IoVs allow cars to communicate with each other and with the surrounding infrastructure, creating a seamless and efficient transportation system. Autonomous driving takes this one step further, allowing cars to operate without human intervention, potentially reducing accidents and increasing efficiency. The smart cockpit is the third field of development, which focuses on creating a comfortable and intuitive user experience inside the car.

Ultimately, the goal of smart vehicle development is to provide better services to users and transform cars into a third living space, in addition to our homes and workplaces. User behavior and preferences are key to achieving this goal, as they determine the final form of smart vehicles. Developers must take into account user feedback and interaction during development to improve user acceptance and satisfaction and create a high-quality experience. Smart vehicles can also be upgraded according to user behavior, which means that they can evolve over time to better meet the needs of users. By providing safe, environmentally friendly, and comfortable transportation, smart vehicles have the potential to transform the way we think about mobility and make our lives easier and more enjoyable.

6.3 Co-robot-driven smart manufacturing plant and process

Cobots, short for "collaborative robots," have been designed to work alongside human employees without causing harm. They manipulate objects in collaboration with a human operator, using virtual surfaces to guide motion. Unlike traditional industrial robots, cobots feature sensors and collision avoidance technology, making them safe to work with humans. Cobots are passive devices, making them suitable for safety-critical tasks like surgery or tasks with large interaction forces like automobile assembly. They are often used in processes that cannot be fully automated, which is why they are frequently seen working in close proximity to humans. Overall, collaborative robotics, which includes robots and robot-like devices, is an essential part of many industries [14–17].

6.3.1 Human–robot collaboration

Human–robot collaboration (HRC) is becoming more prevalent in academic institutions, factories, homes, offices, restaurants, and hospitals. HRC requires interdisciplinary research and development in robotics, artificial intelligence (AI), and psychology due to the complexity of the architecture. AI is especially vital due to humans being a variable factor in HRC. Workers and cobots collaborate on the same job, holding everyone accountable to a common purpose and goal. HRC is like teamwork, with humans and robots committed to finishing tasks through collaboration.

Effective collaboration between humans and robots requires a shared plan and understanding of intentions and actions. Cobots can plan joint movements and work toward a common goal by utilizing saved data and perceiving their environment, resulting in a complete cognition circle. The flow map in Fig. 6–6 shows how outside sensor technologies can enhance cognitive gains, reflecting the potential for AI applications in cobots. This chapter will demonstrate the mechanism and significance of this in collaborative robotics.

Collaborative robot cognition starts with action data being perceived by the system, generating two paths. If the action is recognized, the system gives joint intention estimation and executes the plan. If not, the data are saved into the knowledge base and used for future estimation and planning. The knowledge center informs the operation center, and the loop is closed by joint action.

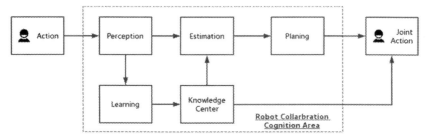

Figure 6−6 Collaborative robot cognition process.

6.3.1.1 The difference between cobot and industry robot

Cobots assist human workers, while industrial robots replace them for heavy-duty tasks. Cobots are easier to operate, learn quickly, and are driving the integration of AI and vision technologies, resulting in high return on investment (ROI). Although cobots have higher hardware deployment costs, additional hardware integration for conventional robots is more expensive. However, a shortage of professional workforce and lack of government regulation may impact cobot adoption. Despite challenges, the market is expected to grow due to AI and machine vision advancements. Some cobot companies are also developing heavy-duty applications.

6.3.1.2 The level of collaboration

Human and robot collaboration is diversified from independent work relation, and sequential collaboration to responsive cooperation. According to the research from the IFR (International Federation of Robotics), the most common collaborative robot applications are coshared work applications where robots and employees work alongside each other, completing tasks sequentially. Often, the robot performs tasks that are either tedious or nonergonomic from lifting heavy parts to performing repetitive tasks such as tightening screws. Applications in which the robot work synchronized with workers in real-time, for example, altering the angle of the gripper to match the angle at which a worker presents a part, are technically challenging. Since the robot needs to adjust to the motion of the worker, and the motion is not predictable, therefore the worker must trust that the full parameters of its potential scope of motion meet safety requirements. The HRC level can be defined by the relation between them. Fig. 6−7 demonstrates the rank of four levels of HRC. The first level represents most of the contemporary industrial robotic

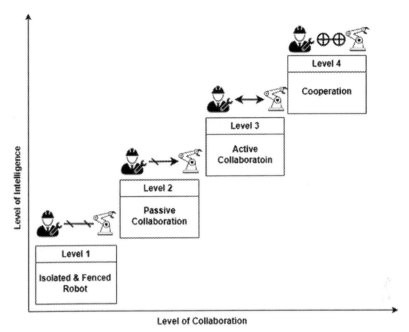

Figure 6−7 The different levels of collaboration in human−robot collaboration (HRC).

automation, that is, human workers are isolated from industrial robots with fences and barriers. Most collaborative robot applications are positioned at the second level. The robots are existing in the same space as human workers and they conduct an action only if their human counter partner pushes the button. At this level, the reason why humans and robots coexist is the development of protective sensor technology.

Strict joint torque and motion speed requirements make unexpected motion a risk for system breakdown. Levels 1 and 2 of HRC are dominant in industrial applications, with one-way collaboration and separated humans and robots. Levels 3 and 4 represent future HRC, where humans and robots form a solid team. In level 3, collaborative robots actively recognize and respond to tasks, boosting production. With AI and cognitive technology development, robots can learn and perform at the same level as humans, working in hazardous environments and becoming part of a team.

6.3.1.3 Benefits of collaborative robots automation

The advent of technologies, such as smart manufacturing, the IoT, artificial intelligence, and edge computing, will revolutionize the traditional

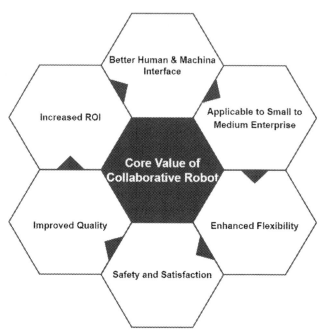

Figure 6—8 Six core values of collaborative robot.

manufacturing system. The adoption of collaborative robots is a powerful example to show how advanced technology can empower enterprises to improve their ROI, quality of employees' safety and satisfaction. Fig. 6—8 shows six important factors covering the economy, production, and societal influence, consist the core value of the collaborative robots.

6.3.1.4 Increased ROI

The economical entry-level of collaboration robots is considerably lower than conventional industrial robots, which makes the collaborative robotic automation feasible for many players. Cobots can be conveniently relocated whenever required. Collaborative robot automation does not require a big change for the rest of the manufacturing process and it will help especially the companies which do not yet have automated production. It will benefit the SMEs (small-to-medium enterprises) and increase their production efficiency. This makes it easy to eliminate any nonproductive activities during working hours. In addition to being highly efficient and flexible, cobots ensure increased ROI due to significantly reduced labor and maintenance costs. This also results in an increased profit margin as well.

6.3.1.5 Better human and machine interface

In any given industry utilizing bots for performing tedious tasks, time, cost, and floor space are the three major critical factors to be considered. This is because the operator can work alongside the cobot, without having to leave the workspace. Cobots are known to reduce the idle time of human workforce by 85%. Consider a traditional assembly line that is set up in a workspace. Here the human workforce sets up the mechanical robots with the required parts to perform the rest of the tasks. The entire production will be bought to a halt for a long time or stopped from its current operations, in case of any required human intervention. While on the other hand, a collaborative robot works along with the human workforce, which increases efficiency significantly.

6.3.1.6 Applicable to small-to-medium enterprises

For some big manufacturers who are depending on the economy of scale, adopting collaborative robots will ease the task of human workers. For example, in the automotive industry, there are some mature automation processes that are already applied by many OEMs, such as white body assembly, robotics welding and painting. At the same time, many repetitive and tedious jobs are still left and they are challenging to be automated by traditional robot integrators. These jobs are still heavily reliant on human labor and the ideal scenario for collaborative robots. In cases like fetching parts and quality inspection, the adoption of cobots will increase efficiency and produce consistent output for a long period of time. The strategies to optimize cost efficiency are a major concern in every OMEs. This is because larger industries, that have a higher production volume, prefer a robot to perform tedious tasks. On the other hand, smaller industries prefer manual labor. Implementing cobots can be beneficial across a range of industry sizes, as these bots do not require a heavy setup process.

6.3.1.7 Flexibility

System integrators play an important role in industrial robot installation because the system experts program the traditional industrial robotic automation and plan the site. Especially for some complex applications, system integration companies need to invest several months with automation experts. The final system is not feasible to re-design or re-program once the previous system integrator withdraws their support. However, for cobot integration, programming interfaces are now becoming increasingly intuitive. For some simple and stand-alone projects, engineers can now easily re-program the cobot to a new task with minimal robot training. The flexibility is ideal for some products having small-to-medium production volume. Cobots are also lightweight and can be relocated to different locations in a factory and they normally take up less space, which is a significant cost reduction for the manufacturers.

6.3.1.8 Safety and satisfaction

Collaborative robots prioritize people, and are designed to free them from repetitive, tedious, and even dangerous tasks. Sensor technology ensures worker safety, reducing risks and improving satisfaction and performance while saving time and money. Collaborative robots are versatile and can be deployed in various industries, including aerospace, automotive, and electronics. Integrating HRC into assembly lines is a complex project but ideal for human—machine cooperation.

The correct robot option, whether we choose traditional industrial robots or collaborative robots, is determined by the real application definition. When speed and absolute precision are the primary automation criteria, it is unlikely that any form of collaborative application will be economically viable. In this case, a traditional, fenced industrial robot is and will remain the preferred choice. If the part being manipulated could be dangerous when in motion, for example, due to sharp edges, some form of fencing will be required. This applies even to cobots that prevent contact. Another factor influencing economic viability is the extent to which the robot must be integrated with other machines in a process. The more system integration needed, the higher the cost of the system installation.

Collaborative industrial robots can help traditional workers to avoid heavy and nonergonomic work, such as holding heavy components. Moreover, cobot can also work on the tedious task, for example, a cobot can pick and place the various screws into the correct position. However, there are still many tasks that are easy for humans but hard to automate, for instance dealing with unsorted parts, and irregular or flexible shapes. Finishing applications such as polishing and grinding applications that require continuous fine-tuning of pressure applied to the surface are also difficult to automate cost-efficiently. Collaborative robotics enables manufacturers to improve productivity by using robots to complement human skills [18].

6.3.2 Discussion

In general, collaborative robots are ideal for SMEs since their application is strictly limited by the return on investment between human workers and labor costs. Collaborative robots can help businesses scale up and automate various parts of production, which additionally frees up more space for remote work. The bottom line is that cobots in manufacturing can realize the benefit to the users, improve quality control, optimize efficiency, and increase production. Furthermore, the current IoT and cloud technologies allow for flexible and versatile use of these machines, that is, the task can be located and optimized during production. The main drawback of cobots in manufacturing is not their functionality cannot meet the industrial demands, but the debate if they are the right choice for a certain application or business. For instance, cobots are not made for heavy-duty tasks, and they cannot handle high-payload jobs. Furthermore, collaborative robots are not fully automated and they are generally playing as an extra hand or an assistant for a worker when they come to the factory floor. More significantly, though, collaborative robots still have some limitations when it comes to complex systems. However, as technology continues to develop, cobots will likely overcome these shortcomings and become a more and more indispensable force in the industry.

6.3.3 The future of cobot

Based on previous market analysis, the cobot industry is expected to reach over 16 billion USD in 2028 [19]. This dramatic increase in market share means cobots are quickly becoming a significant part of automation in manufacturing as the technology is proving to be reliable. There will be more collaborative robot applications emerging in the industry. Cobots will become more sophisticated and versatile with the advance of other corresponding technologies. As AI becomes better, so will the fulfillment of precision and cognitive tasks undertaken by cobots. Moreover, the connectivity of cobots with industrial internet of things (IIoT), meaning the connection to other machines, devices, network databases, etc., will allow them to improve multiple factory workflows and even to provide valuable data analytics, including prediction, analysis, and feedback on how to improve the processes. The ongoing Fourth Industrial Revolution makes the race fiercer and faster, and the use of cobots in manufacturing can give businesses a true competitive advantage. Collaborative robots can augment the work of human employees and allow them to focus on more interesting and less dangerous tasks. At the same time, cobots help to facilitate scalability and increase productivity, efficiency, and output.

With the advent of advanced artificial intelligence, the robotic industry is facing a change: from a programmable robot to a learning robot. This is unlike other AI applications encompassing a large range of data from statistics, mathematics, general science, and economics to philosophy, psychology, neurobiology, and linguistics. Fortunately, robots do not need to address all these areas. The AI technology can provide the robots with problem-solving features: reasoning, planning, learning, verbalization, perception, and manipulation as shown in Fig. 6–9.

Machine learning (ML) is the primary AI technology in the cobot AI system. Machine learning enables cobots to progressively enhance their skills via the experience they gain over time. Cobots can predicate and make their own decision with ML deployment. The perception that appeared before in this chapter is the next ability of a cobot. Large amounts of data are being captured by different sensors around cobots and these data make up the cobot vision system. The significance of perception in the whole cobot AI system involves also safety, which makes the cobot feasible to work alongside their human partners. Verbalization is normally called natural language processing (NLP). All robot companies are investing heavily in this technology. NLP will make robot

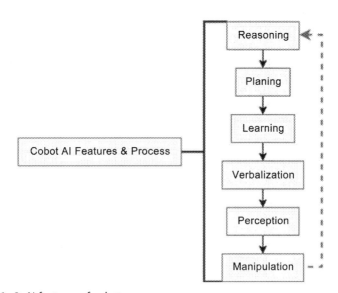

Figure 6–9 AI features of cobot.

deployment easier than now. Through NLP, robotic programming will become convenient and handy. Coding is not necessary when engineers want to change the purpose or relocate the robot to other tasks. Manipulation implemented through AI is a basic skill requirement that allows a cobot to apply a joint motion at its end effector. During the manipulation process, AI is simultaneously performing and learning. Manipulation is a window between physical robotic task and AI.

6.3.4 Cobots in the age of Industry 5.0

With technological advancement and the change in consumer behavior, humans are entering the Industry 5.0 stage. For many years, the production automation system is optimized by digital communication and manufacturers' focus is on how to make a fully automated production and increase production efficiency. This is due to the increase in labor costs and product competition. We normally see this in industry at 4.0 stage and during this stage, AI, the cloud, and IoT improve the data process capability. Change is happening in industry up to the 5.0 stage. In this stage, humans are coming back to center stage. As production technology is refined and developed further, individual value is optimized. Humans are working in the same space with industrial robotic systems. Individual consumption is emphasized. HRC and mass customization are two key terms to describe Industry 5.0 (Fig. 6−10).

In contrast to Industry 4.0, Industry 5.0 aims to put the human touch back into development and production. Industry 5.0 is all about granting human operators the benefits of robots such as technical precision and heavy-lifting capabilities. This ability for humans to perform difficult or laborious tasks with relative ease will allow the introduction of a higher degree of control and the ability to individualize every phase of production [20].

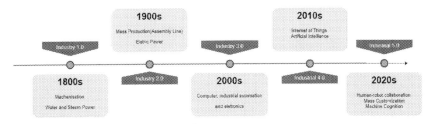

Figure 6−10 From Industry 1.0 to Industry 5.0.

Where Industry 4.0 was focused mainly on ensuring consistency of quality, flow, and data collection, Industry 5.0, while still focused on these goals, put more attention on highly skilled people and robots working side-by-side to create individualized products. Collaborative robots can perform the task safely and manage the risk level when humans are present in the workstation [21]. The future cobot technology will let them understand their target and current work in order to make a plan to finish the job on time. The future cobot will start from basic input from the robot operator and gain the know-how gradually. When they finish their learning curve, they will perform at the same level as their human partners. So, the experience to work alongside a cobot in the Industry 5.0 stage will be authentic and it will increase the human worker's satisfaction as they have authentic interaction with their partner robots. With Industry 5.0, that delineation gets hazier as humans and robots start to work side-by-side and even together. Collaborative robots will accomplish the required tasks of heavy lifting and ensuring consistency while the skilled human provides the cognitive skills of a craftsperson [22].

6.3.5 Conclusion

With technological revolution comes changes in the job market, requiring workers to re-skill and be educated to adapt to labor market demands. Closer collaboration between companies and educational institutes is necessary to equip workers with necessary skills, including analytical, communication, and decision-making. Continuous, technology-based learning programs are crucial for upskilling existing workers. Humans will remain central to manufacturing, but personalization and mass customization will reshape the industry. Flexible, mobile automation methods will become popular, making collaborative robots a good match for the future. Mobile automation offers affordable collaborative robots and automation solutions for various industries.

6.4 Intelligent integrated energy services and society ecosystem

6.4.1 Integrated energy system and intelligent integrated energy services

IES is a new energy service system that takes electricity as the core, integrating gas, heating, transportation, and other energy systems [23].

It relies on the continuous innovation of cogeneration equipment, energy conversion equipment, renewable energy, and other related technologies to achieve a high degree of synergy among multiple energy production, transmission, conversion, storage, consumption, and other aspects.

In the context of the energy Internet, the development of measurement, communication, Internet, and control technologies provides essential technical support for information interaction and cooperative control among heterogeneous energy flow systems such as electricity, natural gas, and heating [24]. As shown in Fig. 6–11, the IIES [25] serves people's needs for a better life and efficient productivity.

IIES involves various aspects of energy supply, transmission, and end-use. Specifically, IIES covers different types of energy businesses, for example, energy sales together with services such as energy infrastructure construction, distributed energy, energy saving and emission reduction, demand response, and financial derivative.

Compared with traditional energy services, IIES is expected to reduce the dependence of human society on traditional fossil energy and promote the consumption of a high proportion of renewable energy. It is an effective carrier to realize the integration of "4 networks 4 flows" [26].

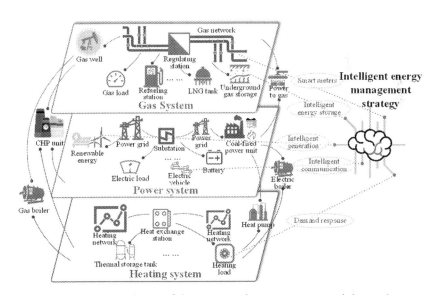

Figure 6–11 Conceptual map of the integrated energy system and the implementation path of smart energy services.

IIES breaks the barriers of heterogeneous energy networks such as electricity, gas, and heat and enriches the way human society uses energy. The most common ones are alternative energy supply or consumption, such as switching between natural gas stoves and electric cookers, multiheat source complementation between gas boilers and heat pumps, and transportation electrification. Meanwhile, a clean energy carrier represented by hydrogen energy has emerged, and the replacement of fossil energy has been gradually realized by employing electric hydrogen production, fuel cells, and HCNG technologies. In addition, new building energy conservation technologies and intelligent energy management strategies can improve the self-sufficiency of the demand side, such as near-zero energy buildings and smart towns that have emerged in recent years.

IIES promotes the benign interaction between the humanities network and the energy network and enhances the identity of human beings in the energy network. Under IIES, human beings have a wealth of energy usage, transportation, and cultural lifestyles. Users equipped with distributed new energy and energy storage can transform from simple energy consumers to prosumers. New energy service providers such as energy storage service providers and load aggregators have even emerged. Humans can adjust their energy use behaviors and methods in response to incentive and punishment signals, such as energy prices and carbon taxes, and interact positively with the energy network, serving the needs of the energy network in terms of peak regulation, frequency regulation, and renewable energy consumption.

6.4.2 Basic characteristics of intelligent integrated energy services

The basic features of IIES can be summarized into three points: multienergy flow, multiagent, and multiscale, which reflect the rich connotation of IIES from different angles [27]. "Multienergy flow" is the fundamental feature that distinguishes IIES from traditional independent power, heating, and natural gas systems. On the one hand, the electricity, gas, and heating subsystems in IIES have their energy service resources (called flexibility resources) and energy service demands (called flexibility demands). Fig. 6—12 sorts out typical IES flexibility resources from the two dimensions of response capacity and response speed, covering equipment and facilities involving multiple energy flows. It can be seen that the specific physical characteristics of energy services in different subsystems may differ

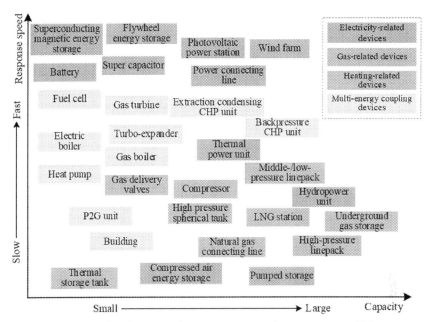

Figure 6−12 Typical multienergy devices in integrated energy system (IES).

to some extent. Specifically, the energy service of the power system is expressed as the system's long-term energy balance and short-term power balance. In contrast, the energy service of the heating system is described as the system's long-term thermal energy supply and demand balance and the short-term heating economy and environmental protection level. Additionally, the synergy and complementarity among multiple energy flows can be regarded as a kind of virtual energy service. To be specific, through the synergy and regulation of multiple energy flows, the flexible resources of electricity, gas, and heating systems can be fully utilized to provide energy services for diversified energy users.

"Multiagent" means that the flexible resources mobilized by IIES involve different stakeholders. Electricity, gas, and heating subsystems usually belong to different departments. Typical core demands of various departments include economic efficiency, green and low-carbon energy supply, and energy supply safety and adequacy. The improvement of certain performances of one entity may be at the expense of other entities. Therefore IIES between different entities is accompanied by the transfer of costs. The flexible resources of each link may belong to independent service providers, such as energy storage service providers and load aggregators.

Then, flexible resources become a unique commodity in the energy system, and the realization of its value depends on an effective market mechanism. To sum up, the "multiagent" feature means that some flexible resources in the IIES are difficult to allocate in a unified manner, and a suitable market mechanism is needed to explore their potential. In addition, the "multiagent" feature also brings problems such as information barriers.

"Multiscale" refers to the multitime-scale and multispatial-scale features of the IIES. On the time scale, energy demand has a different performance in a year, quarter, month, day, and real-time. The focus is gradually shifted from the mid- and long-term regulation capability of system energy to the short-term or real-time regulation capability of system power. On the spatial scale, according to different application requirements, IIES cannot only focus on various building users (such as energy-saving and fee-reducing services of the smart home) but also can be expressed as city-level and even regional-level service capabilities (such as regional demand response provided by aggregators for external grids).

6.4.3 The basic theory and methods for intelligent integrated energy services

The core technical support of IIES can be summarized into three aspects: modeling and simulation, planning and design, and operation optimization. Modeling and simulation are the basis of the operation, planning, and analysis of IES. IES modeling can be further divided into equipment modeling and network modeling. Equipment modeling includes models of generators, various prime movers, and energy storage systems in power systems, models of gas transmission stations, pressure regulating stations, and gas storage systems in natural gas systems, as well as models of heat sources, heat exchangers/heat exchange stations in heating systems. Network modeling involves power distribution and transmission, gas distribution and transmission, and heating supply networks. The IES simulation can be divided into three parts: static energy flow calculations, dynamic energy flow calculations, and dynamic simulation analysis considering the dynamic characteristics of the equipment. The purpose of the static energy flow calculation is to obtain the system's operating parameters at a certain time period according to the load demand and the output of the source node to provide a basis for the safe and economical operation of the energy system. Based on static energy flow, dynamic energy flow calculations further consider the unique delay characteristics of energy networks such as heat and gas grids in the energy transmission process and obtains the changing laws of parameters such as power and flow over time. Based on the dynamic energy flow calculation, the dynamic characteristics of the system components are further considered, which constitutes the dynamic simulation analysis of IES.

Unlike the traditional energy system's planning of single energy forms such as electricity, gas, and heating, the planning and design of IES needs to break down industry barriers, remove policy and geographical restrictions, and give full play to the complementary advantages of energy sources. Scientific and reasonable IES planning can meet the needs of the planning target in various aspects, such as economic benefits, renewable energy consumption, energy saving, and emission reduction. IES planning needs to consider the resource endowment, economy, policy, society, environment, and other factors in the planning area. The designed equipment-level and network-level power, gas, and heating subsystems must meet the planning area's multienergy load demand. Stochastic programming, robust programming, multiobjective programming, and multistage programming are the commonly used IES planning methods.

IES operation optimization aims to determine the optimal output of the equipment in the system and the optimal power distribution of the network to optimize the system operation economy, environmental protection, or energy efficiency. For a given load profile, the system often has a variety of feasible operation schemes, but different schemes may have significant differences in performance. Therefore operation optimization is the key to exerting the efficiency of IES. From a physical point of view, the energy flow of electricity, gas, heating, and other energy flows in an IES presents a complex coupling relationship through energy conversion equipment. The redundant configuration of the system and the complementary characteristics of various energy flows bring relatively high operational flexibility. In contrast, factors such as renewable energy output, multienergy load, and meteorological factors that are difficult to predict make the system operation face multiple uncertainties. To summarize, IES operation control is a complex problem with multienergy flow coupling, multitime scales, and multioperating objectives.

6.4.4 Typical intelligent integrated energy services platforms

In terms of the IIES platform related to modeling and simulation, the separate power, gas, and heating simulation modules have developed rapidly. The mature commercial software includes TRNSYS, Modelica, and

TERMIS, which can perform heat network simulation; the DIgSILENT/ PowerFactory, pandapower, PSSE, and DGRSS, which can perform power simulation; and the Pipeline Studio, which can simulate gas systems [28]. However, commercial software to support integrated energy cosimulation has not been effectively developed. The TRNSYS only supports the simulation of a limited number of devices and networks in electrical, cooling, and heating networks. In 2020 Southeast University developed the IES-Sim, a simulation platform for electricity-gas-cooling-heating-steam IES equipped with more than 50 devices and 5 network models, but it has not yet been commercialized. The operation interface of IES-Sim is presented in Fig. 6–13. Users can place, drag, and connect multienergy equipment and pipelines on the virtual canvas, and set their technical parameters. After completing the configuration, users can enter the power flow calculation or simulation function. A typical example of a three-park electricity-gas-heating-steam IES is shown in Fig. 6–14. The power flow calculation and dynamic simulation results are given in Fig. 6–15.

In terms of IIES platforms related to planning and design, several platforms, including the HOMER, DER-CAM, CloudPSS-IESLab, and IES-Plan, have been developed at home and abroad in recent years [29]. DER-CAM and HOMER were developed by Lawrence Berkeley National Laboratory and National Renewable Energy Laboratory (NREL), respectively. The CloudPSS-IESLab and IES-Plan were developed by Tsinghua University and Southeast University, respectively. Among them, the DER-CAM and HOMER focus more on the planning and evaluation

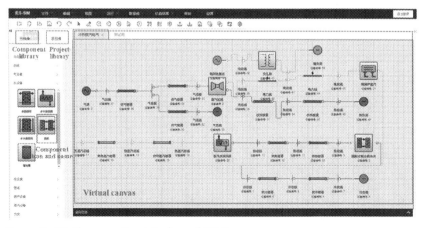

Figure 6–13 The operation interface of IES-Sim.

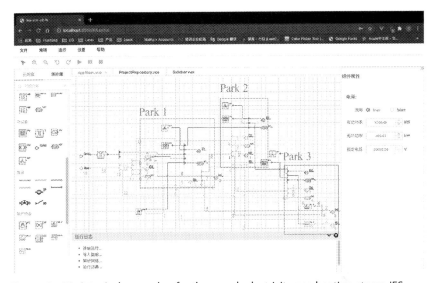

Figure 6–14 A typical example of a three-park electricity-gas-heating-steam IES.

of power systems, while the CloudPSS-IESLab and IES-Plan consider the coupled planning and simulation of multiple energy forms and multiple integrated energy equipment. The platform architecture, main planning process and core functions of IES-Plan are shown in Fig. 6–16.

Regarding the IIES platform related to operation optimization, the relatively mature IES energy management platform at home and abroad is the IEMS developed by Tsinghua University in 2020 [30]. In general, the above theoretical research and platform development is committed to providing better IIES. Future work will be carried out for the following aspects: improving the modeling accuracy of equipment and networks in IES, emphasizing the role of users in participating in intelligent energy services as prosumers, considering the relationship between multiple departments coupled by multienergy flows, and quantifying the interaction between energy systems and the external environment.

6.5 Digital Renaissance-driven smart city

Four network four flows (4N4F) is the foundation of smart cities, which have evolved with its development. Urbanization brought about

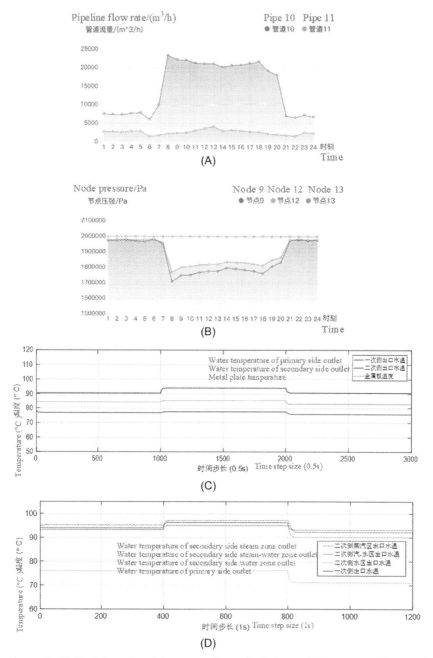

Figure 6—15 Partial results of the power flow calculation and dynamic simulation of the example: (A) pipeline flow; (B) node pressure; (C) water—water heat exchanger; (D) steam—water heat exchanger. Please check the online version to view the color image of the figure.

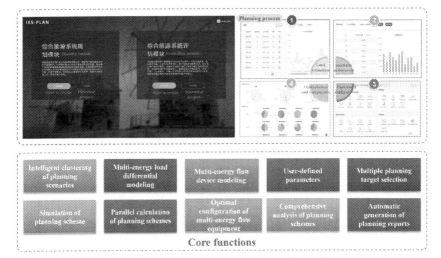

Figure 6–16 The planning process and core functions of IES-Plan.

"urban diseases," such as population density and space scale, requiring "smart solutions." Smart cities have undergone four stages with different processes, technical characteristics, and updated concepts, contents, and applications. They involve many issues, including government policy, infrastructure, and social culture. This section will focus on the historical origin, technological relevance, application cases, and philosophical thinking of smart cities [31,32].

6.5.1 The emergence of smart city

The city is the concentrated embodiment of human civilization, recording the social changes and historical precipitation of human production and life. There are four stages of human societies with diversified urban functions, layout and scale: primitive nomadic, agricultural, industrial, and information societies. The speeding process of urbanization is the fundamental reason for the emergence of smart cities.

As shown in Fig. 6–17(A), thanks to the steam engine of the First Industrial Revolution, factories could enjoy the production mode of assembly line which attracted a large number of agricultural people to cities. The urban population density increased and transportation technology developed sharply, all of which led to traffic congestion and industrial pollution. The transportation network formed since the transportation intensification brought by train, steamship, and car, and "material flow"

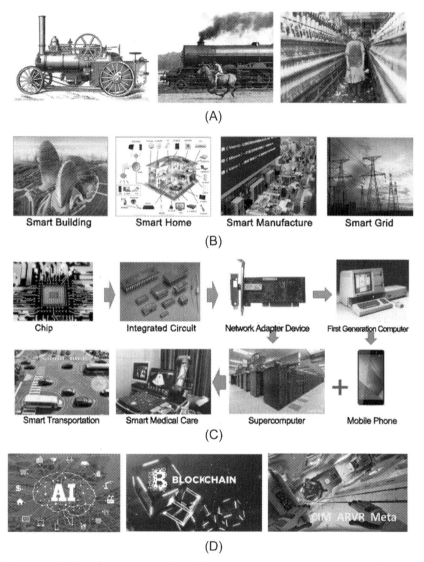

Figure 6−17 The four network four flows (4N4F) and smart city. (A) The First Industrial Revolution—transportation network and material flow. (B) The Second Industrial Revolution—energy network and flow. (C) The Third Industrial Revolution—information network and flow. (D) The Fourth Industrial Revolution: humanity network and value flow.

became the main application and modernization embodiment of the first-generation smart city. As shown in Fig. 6−17(B), power generation proclaims the arrival of the electrical age. Energy progress has improved the coal pollution problem, and the "energy network" formed with the interconnection of oil, natural gas, water conservancy, wind power, nuclear power, and others, which drove the emergence of electrified smart cities. The "baby boom" after the Second World War expanded the urban population rapidly, and the need for larger spaces to live more comfortably made intelligent office buildings, electrical automation shopping malls and smart homes became the main characteristics of the second generation of smart cities.

As shown in Fig. 6−17(C), the first computer marks the arrival of the Third Industrial Revolution. The invention of semiconductor integrated circuits, broadband, mobile network communication products, smart sensing and network control equipment realized the informatization and networking transformation and also speeded the arrival of the era of "Internet + " with social digitalization, information network, and intelligent management. "Information flow" is the key element of urban wisdom and soul resources endowed smart cities with connectivity. As shown in Fig. 6−17(D), the Fourth Industrial Revolution has profoundly changed the competitive rules and comparative advantages of countries and cities, thus exerting a profound impact on the global landscape. Countries and cities around the world have made their own digital development strategies, listing smart cities at the top. Since 2010, the Smart City action plan has been carried out on a large scale at home and abroad. For example, the EU announced the "Smart Cities and Communities Innovation Partnership Strategy Implementation Plan" in 2013 [33], India launched "Digital India" in 2015, the EU issued "Digitizing European Industry Initiative, DEI" in 2016, Russia officially approved the Digital Economy Program of the Russian Federation in 2017, and the United States released "A U.S. Grand Strategy for the Global Digital Economy" in 2021.

UNESCO put forward the concept of an eco-city in the 1970s [34], and IBM first raised the concept of the smart city in 2008. Countries around the world therewith put forward smart city development goals and strategic plans, such as the European Union's "Smart Cities and Communities European Innovation Partnership, SCC-EIP," Japan's "I-Japan Strategy," South Korea's "U-City Plan," Singapore's "Smart Nation Platform," China's "Smart City Demonstration Projects," and so on.

In 1998 then US Vice President Al Gore put forward the Digital Earth strategy, marking the starting point of the fourth digital Industrial

Revolution. The digital Industrial Revolution, different from the previous three industrial revolutions, has constructed a new economic era of data intelligence, and some scholars suggested it as the fourth industry [35]. Based on digital productivities, the digital revolution set up new digital production relationships such as virtual reality (RA), digital city model CIM, and chain blocks on the global geography and ecology, economy and social operation, humanities and human settlement, which greatly liberated the human brain and brought the information explosion. And artificial intelligence era is the latest stage of Digitalization marked by big data, high computing power, and large mode. If the First Industrial Revolution liberated human physical strength, the second provided machine power, the third released information dissemination, and then the fourth expanded the space of the human brain and consciousness. IESE Cities in Motion Index 2020 (seventh edition) offers a series of conclusions and recommendations with "people first" topping the list, which means satisficing the consciousness for a better city and a smarter life matters most. The value flow has become a fundamental driving force for city's future development.

Technology progress is the original driving force of the smart city. The technical characteristics of smart cities correspond to the "fashion themes" of the Industrial Revolution in different periods, and industrial revolutions are reflected by the intelligence traces and material precipitation of urban infrastructure, economic investment, and social models. Smart cities, experiencing four development stages: automation, intelligence, networking, and intelligence, correspond to four network four flows (4N4F) in different periods: the steam engine revolution corresponding to the transportation network, the electrical revolution to the energy network, the information revolution to the information network, and the digitalization to humanity network. Details can be seen in Fig. 6−18.

Four industrial revolutions are not substituted but superposed. The achievements of each Industrial Revolution have added and developed to the present day. The different characteristics of smart cities at four stages also cannot be substituted, but cross-promoted and integrated. Therefore 4N4F, vertically deepening and horizontally integrating, could represent the development direction and characteristics of human material civilization, technical knowledge, and way of life. The transportation, energy, information, and value flows are the "living" element resources for the operation of smart cities.

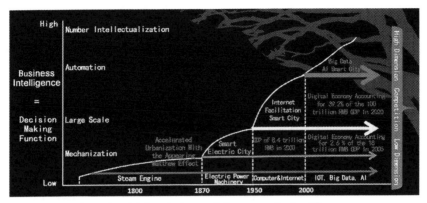

Figure 6−18 The Industrial Revolution and the smart city in different stages.

6.5.1.1 The networking transition of urban layout

The development of a smart city corresponds to the emergence of four networks, reflecting the change in urban layout. In ancient times, the urban layout was basically divided by mountains and water and organized into clusters to meet the requirements of management and defense security. It was based on physical geography and human activities. The gravity model is the earliest urban morphology study. In 1858 according to Newton's law of universal gravitation, Cali put forward the formula that the attraction between cities is proportional to the city scale and inversely proportional to the distance. In 1931 W. J. Reilly further developed the formula and optimized settings of city functional zoning, scale, and distance. The layout of traditional cities, mostly around religious or centralized architecture, was symmetrical and radioactive, and surrounded by walls. According to Matthew's effect of urban gravity [36], constantly expanding made large cities and megacities become stronger and stronger and face the brink of collapse. Urban diseases such as traffic congestion, environmental pollution, the heat island effect, and rising living costs waited to be addressed urgently.

Thanks to the transportation network in the First Industrial Revolution, the urban layout was structured by the road network. City was divided into blocks as the basic units, and was planned, built, and operated by zoning (divided districts into blocks of 50−100 m). The intelligent transportation network not only included the intelligent assembly of transportation facilities, but also the transformation of the facilities both for road and car. Based on the road network, the municipal pipe network system, the ground street lighting and monitoring network systems, and the block administrative management grid was also constructed. Therefore the transportation network becomes the neural network of urban layout at that time. The energy network in the Second Industrial Revolution changed the urban layout. Power station, substation, power grid, and dispatching system, as new elements of municipal support, accelerated the formation of multifunctional urban networks. The power grid entered all buildings and families, this expanded coverage of urban networks and intelligent control, while the birth of "the sleepless city" has extended the time of urban production and life. The brightness of urban night lighting seen from satellites reflected the activity of urban economic prosperity and urban electrical intelligence.

The information network of the Third Industrial Revolution realized the urban network extension—"the world is flat," and cities are members of the "global village." Information facilities become the new urban infrastructure, including 5G, IoT, IDC, Cloud, and AI. On the basis of the urban network formed by the first two industrial revolutions, the information network, together with the Internet based on TCP/IP, the wireless communication network, and the mobile network integrated air and space, the city was built into a ubiquitous network system that can maintain dynamic real-time broadband links at anywhere and anytime. This kind of high-precision, high-resolution, and full-coverage perceptual network laid the foundation for the comprehensive digital network intelligence.

There are four layers of the network system for the modern smart city, namely: sensing, data or knowledge, and application layers. Its core used the new generation of information technology to change the interacting way the government, enterprises, and people adopted, thus improving the clarity, efficiency, flexibility, and response speed of interaction, so as to achieve the perfect combination of the information infrastructure and the highly integrated basic infrastructure. The network layer was the most intuitive embodiment of integration and coordination in the smart city framework model. It was also the neural network and central core of the smart city intensive sharing, as well as the basis of smart city operation and management. Therefore the networking change of urban layout reflected the biggest change of urban modernization since the Industrial Revolution, as shown in Fig. 6−19.

6.5.1.2 Digital trend of modern and future city

With the arrival of the knowledge economy, the world has entered the era of big data and artificial intelligence. According to experts, with the

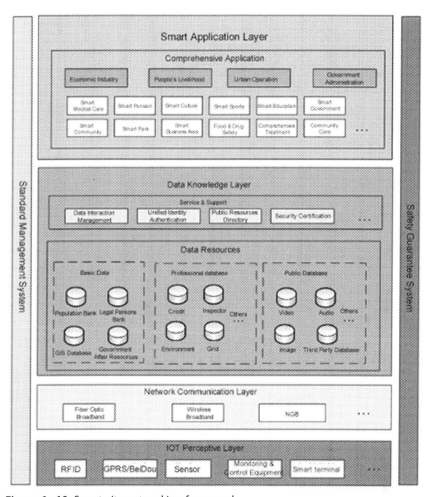

Figure 6—19 Smart city networking framework.

popularization of robots, 80% of manpower in the traditional industrial, agriculture, mass transportation, urban construction, and operation management can be saved and converted into creators of digital twin space thought flow and consumers. The humanistic relations network corresponding to digital society formed the new value and production relations, among which some rules and fields were linked with the real world, and others were just human consciousness and value desire. The metaverse, twin cities, sci-fi cities, and game spaces have raised the curtain of the future urban form.

Digital twin city is a new urban form reflecting human consciousness, value, and default rules to the physical one, which aims to the better management. It is an extremely high concurrency city-level information infrastructure. Based on the city information models BIM and CIM, all dimension data can be acquired by satellites, airborne remote sensing mapping, and ground sensors, to form the dynamic urban spatiotemporal big data with whole domain, total factor, and high precision, so as to realize fully digital mapping. Thus a more intelligent and effective resource platform across traditional government affairs, economic operations, and social management can be achieved. It provides a perfect urban digital base for the city to realize self-learning improvement, virtual-real intelligent interaction, and people-oriented decision-making. The metaverse, based on digital twin technology, creates digital content to form a new and transformed digital world. In the metaverse, people and machines can be supported to continuously create and generate value. People can transform at will and give play to their ideas, and then generate value through content operation. As a portal of digital experience, the metaverse is a value platform of production, transportation, and consumption with huge energy, which has a big effect on the real economy and will constitute the operation ecology of the future city.

The rapid development of digital technology is reshaping our society in an unprecedented degree. Machines will also have lives, and digital technology can make them self-revolve like people. Therefore new values and humanity are needed to interact and regulate all of these evolutions. In that condition, human understanding is the key to developing the breakthrough leading technology. We should explore the new area decisively but with caution and optimism.

6.5.2 Smart city: the integration and innovation of four network four flows

To further study the theoretical innovation of 4N4F, we should explore it from philosophy. This section will discuss the philosophical view on the urban development origins, as well as the productivity factors and production relations promoting the smart city reform.

The author of *The City in History* [37] proposed that urban development is the presentation of the human civilization evolution in the material space. It has three characteristics: magnet (attraction), container (carrier space), and evolution (constantly evolving and developing with human consciousness). 4N4F is the reflection of the urban characteristics in different historical stages, as well as the technological productivity and production relation innovation promoting the city development since the Industrial Revolution. Cities are the most direct, concentrated, and large-scale carriers of these innovations.

6.5.2.1 Four flows: the most active productivity factors

The 4N4F are summarized from innovation productivity and production resource factors, as well as the organization mode innovation of production relations brought by them since the Industrial Revolution. In this section, we discuss and illustrate it from the smart city's evolution and its internal relations.

In the First Industrial Revolution, steam engines brought cars and rail transportation. Thanks to that, factories moved into cities on a large scale, but "urban diseases" began also because of the traffic congestion and air pollution they brought. In the early 19th century, the automobile popularization made people want to live in nice suburbs and remote villages and commute to cities. Large numbers of wealthy people and residents have left urban centers, resulting in the hollowing out of them and worsening pollution and traffic congestion. Facing the dilemma of the "automobile" era, some scholars, including Sam Casella, former president of the American Planning Association, put forward the Smart Growth Theory, which required people return to the city, construct an intensive transportation system, and reconstruct the CBD function of the central city.

Energy flow is the power flow of modern city operation. Without it, electrical automation, network intelligence, and artificial intelligence cannot operate successfully. Therefore one of the key factors to enable smart cities is to realize the intelligent and low-carbon energy flow: to generate the best operating state and ecological balance with the minimum energy flow consumption, this is a challenge and also an opportunity. At present, many smart cities around the world have set "double carbon" standards as key goals for 2025−60.

The proposal of humanity network and ideological value flow is obviously different from the traditional four flows and more challenging. Some scholars predict that with the realization of digital economy and smart city, 80% of the existing production and social management personnel will be gradually replaced by ubiquitous IoT robots, big data, and artificial intelligence. The metaverse, digital twin city, and online virtual world will be the new space to experience exchange of consciousness and value, online business, and life consumption. The digital world can correspond to the material world, and it can also be completely virtual, realizing the liberation of human intelligence. In the discussion of idea flow and consciousness flow, some scholars also put forward the concept of heart flow [38]. *Social Physics: How Good Ideas spread-the Lessons from a New Science* written by Alex Pentland of the United States, proposed the idea flow [39]. In *A Brief History of Mankind*, Yuval Harari of Israel discussed that the ability of language and thought are the most fundamental elements of the development of human wisdom.

Just as it put forward in *The City in History*: the achievement and direction of urban development are the result of human consciousness. The value flow or consciousness flow, the core of the element flow in the humanity network, reflects the value orientation of people, and is also the base to construct the people-oriented digital world.

6.5.2.2 Four networks: new operation system of smart city

The four flows are elements corresponding to the productive forces of four industrial revolutions, while the four networks are the carriers of the four flows. The current urban economic and social operation has been systematized by the networking supply chain in the vertical direction, and unitized by the grid in the horizontal direction. From road networking, energy networking to information networking, social organizations and individuals become more efficient in collaboration, low-consumption transmission, and accurate matching, so as to provide a space spanning time and space for the humanity relations networking. The network must have the actual function so as to keep sustainable long-term operation. The former US Vice President Al Gore put forward the information superhighway in the last century and the strategy of digital Earth at the same time. The infrastructure and technological progress of the information superhighway, with its content of digital and information flow, formed the rapid development and popularization of industrialization, socialization, and globalization. Therefore the four networks and four flows must be a unified whole, and the basic composition of their own philosophical cognition.

In the process of urban intelligent development, the four flows demand promoted the emergence of the four networks and constructed the new production relationships and operational mode of the city. First, the transportation flow has changed the urban grouping and natural segmentation forms, the road grid has been born into the block and administrative division; elevated roads, subway, and other underground channels

have expanded the urban spatial layout; as well as vehicle-road cooperation and intelligent driving have become the new urban embryonic forms
of business. The power grid and its facilities are the most important infrastructure of the city. The intelligent transformation of the urban power
grid and the completion of power supply supporting facilities have
become the biggest challenge and ecological goal of urban operation,
energy conservation, and emission reduction. The requirements of information flow on information networks made more new elements such as
urban brain, IDC, IoT, Cloud, and AI integrated into urban infrastructure. And with the emergence of digital twin cities and urban model
(CIM), large-scale dynamic tracking, fine realization of digital economy,
and smart city, 80% of the existing production and social management
personnel will be gradually replaced by ubiquitous IoT robots, big data,
and artificial intelligence. The metaverse, digital twin city, and online virtual world will be the new space to experience exchange of consciousness
and value, online business, and life consumption. The digital world can
correspond to the material world, and it can also be completely virtual,
realizing the liberation of human intelligence. In the discussion of idea
flow and consciousness flow, some scholars also put forward the concept
of heart flow [38]. *Social Physics: How Good Ideas spread-the Lessons from a
New Science* written by Alex Pentland of the United States, proposed the
idea flow [39]. In *A Brief History of Mankind*, Yuval Harari of Israel discussed that the ability of language and thought are the most fundamental
elements of the development of human wisdom.

Just as it put forward in *The City in History*: the achievement and
direction of urban development are the result of human consciousness.
The value flow or consciousness flow, the core of the element flow in the
humanity network, reflects the value orientation of people, and is also the
base to construct the people-oriented digital world.

6.5.3 Cases and practices

Smart city construction, inlcuding that of digital government, digital
economy, digital society, digital culture, and digital ecology, is a long-
term and global topic. At present, more than 1000 smart cities have been
launched or are under construction in the world. According to the
research of Technavio [40], the market share of smart cities is expected to
increase to 151.99 billion dollars from 2020 to 2025, with a compound
annual growth rate of 19.43%. Depending on the history background,

natural environment, geographical positioning, population size, economic form, and many other factors, each smart city has its own characteristics. Thousands of smart city cases are being constantly updated. Combined with the development path and characteristics of 4N4F, eight typical cases are introduced in this section to give readers a deeper understanding.

6.5.3.1 Top-level designs and policies

The overall planning of Xiongan New Area deployed by the central government targets on realizing all-round smart based on the urban information model. The professional data are integrated on the digital twin platform so as to achieve a new pattern, leading the smart city to a new stage of digital twin. Three new cities—ground, underground, and digital—are planned and will be implemented synchronously to create the digital twin city. The Xiongan Urban Computing Center (Supercomputing Cloud), also known as the "smart Brain" of Xiongan, will provide network, computing, storage, and other services for big data, blockchain, IoT, AI, VR/AR, and meet the data combing needs of Xiongan smart city in different scenarios.

6.5.3.2 Basic infrastructure: base of digital city

In recent years, the new infrastructure such as IoT, cloud computing, big data, artificial intelligence, and AI has been highly valued. The central and local governments issued a number of policies and plans to promote the development of new infrastructure. Singapore, as the representative, formulated the 10-year development plan (iN2015) for the information and communication industry to enhance the global competitiveness, and cooperated with the Sensible City Lab of MIT to build urban big data platform called Live Singapore [41].

The Urban Information Model (CIM) basic platforms were fully applied in Guangzhou to continuously promote the construction of digital twin cities and new smart cities, making urban governance more refined. An app called "Suzhiguan" has connected 35 municipal departments with 115 business systems, joined up 84,000 IoT sensing terminal devices, collected more than 7.2 billion urban operational data items, generated 3103 urban physical signs data items, and constructed five panoramic views of basic urban elements of "people, enterprises, land, property and government." It has realized the "one-screen with all views" of urban operation situation and the "one-map with all collection" of key indicators of urban operation signs, providing intelligent support for leaders to make scientific decisions. All of these are inseparable from the digital base like CIM.

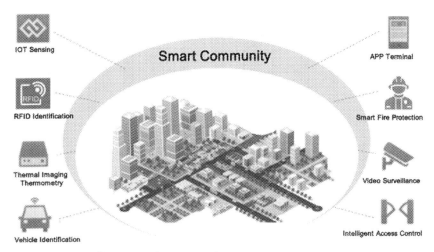

IOT Sensing

Smart Community

APP Terminal

RFID Identification

Smart Fire Protection

Thermal Imaging Thermometry

Video Surveillance

Vehicle Identification

Intelligent Access Control

Figure 6−20 Intelligent security community.

Community is the foundation of smart city construction. Smart city, as a collection of smart communities, represents the wisdom concept at the city level. The "wisdom" of the smart community is reflected on three aspects: energy management, smart security, and smart service. For example, the Kitakyushu smart community in Japan [42] is a representative of the citizen participatory smart community that pays attention to regional energy management, while Fujisawa Sustainable Smart Town is dedicated to elderly care services, carbon emission and domestic water reduction, earthquake emergency protection, among other things. In China, Chengdu has built 1149 provincial smart security communities, in addition, Chongqing Municipal Commission of Housing and Urban-Rural Construction issued a plan in 2021 to encourage real estate enterprises to build two-star and above smart communities, reflecting the increasing importance of intelligence in Chinese construction industry (Fig. 6−20).

6.5.3.3 The construction of urban brain

Hangzhou became the first city in China to explore the urban brain in 2016. The urban brain is composed of a main center, system platform, digital cockpit, and scene. It has launched 48 scenarios and 155 digital cockpit applications in 11 fields, including traffic, police, and health. Hangzhou City Brain has achieved a national lead in implementing the "rod-free parking lot" and the "digital park card" first, as well as realizing "20 seconds to enter the park, 30 seconds to check in," and "all pay once after medical treatment."

6.5.3.4 Great activity to be a demonstration

The 2022 Beijing Winter Olympic Game showcased a classic application of smart cities, integrating technological achievements such as digital twin, AI, VR, smart healthcare, and smart transportation. Robots were used for epidemic prevention, catering, and service. Wearable smart thermometers and high-tech products provided guarantees for epidemic prevention. The "Eagle Eye" assistance system and 4K track camera system "Cheetah" ensured fair competition. The event was the first Olympics to put the core system on the cloud, migrating all information to the cloud for a historic project.

6.5.3.5 Smart transportation in constructing hub city

Congestion tax, automatic tolling system with laser, camera, and engineering technology, as well as the basic infrastructure of three-dimensional traffic network and dispatching system, successfully built Stockholm to be the first model city of "smart transportation" in the world [43]. Paris ranked third among the world's top 10 smart cities in 2012, thanks to the smart billboards and electric car sharing.

Beijing Capital International Airport has realized the digitalization of all scenes and processes, collecting more than 10 million aircraft location data, 20 million vehicle location data, and nearly 100 million environmental monitoring data detected by 23,000 sensors every day in the digital twin world to support intelligent management. Besides, the Cheng-Yi high-speed intelligent system on Chengdu to Yibin in Sichuan achieved centimeter-level accuracy, with smart poles set up every 800 m and new millimeter wave radar equipment for 24-hour real-time perception. China's key transportation project equipped 2754 km of the Yangtze River with over 5200 beacon telemetry remote control terminals, 155 automatic water stations, 219 waterway maintenance vessels, and 34 traffic signal stations, providing important digital waterway data. The system uses Beidou, 5G, and IoT technologies to transmit real-time data to the dynamic monitoring system, displaying the main waterway of the Yangtze River in three dimensions.

6.5.3.6 Smart energy in building green ecological city

The smart grid was first proposed by the United States [44], but China developed fast and has reached the world's most advanced level in both hardware and software. In 2018 more than 80 people were required to inspect a 3000 km power grid in Shenzhen, China, while in 2021, with the "video + UAV + artificial intelligence" intelligent transmission inspection system, most of work could be completed automatically. Now, more than 3000 smart cameras have been installed on power lines throughout Shenzhen, screening out abnormal situations from more than 40,000 real-time snapshots every day, the failure rate decreased more than 30%. The equipment failures can be determined in almost half an hour, which greatly improves the efficiency and energy saving.

The British scholar Edward first proposed the concept of "garden city," drawing attention to the integration of urban development and ecological environment. Following this, Japan implemented the Green Tokyo University plan in June 2008, and in 2010 selected Toyota City to vigorously develop the construction of low-carbon cities. By 2019, a grid monitoring system for air pollution control was established, in the streets of Huangpu District in Guangzhou (including six air index, TVOC, and particulate matter detectors). Air quality monitoring networks, Internet, big data, satellite remote sensing, sensors, and artificial intelligence are being applied to environmental monitoring. The large-scale monitoring data from remote sensing satellites and basic monitoring data from ground stations together realize the three-dimensional monitoring and comprehensive analysis to the ecological environment.

6.5.3.7 Ubiquitous perception cities with information networks

The development of the Internet has gone through Web1.0, Web 2.0, and Web3.0. Web1.0 users obtain content only, Web2.0 is a readable and writable Internet where users can generate content; while Web3.0 incorporates new technologies such as big data, AI, blockchain, and metaverse. In the era of Web3.0, the network is everywhere, and it is inseparable from human life. The network has expanded from a partial aspect of human life to a panoramic scene.

In China, WeChat, Taobao, Alipay, and TikTok have become super-platforms with large user scales, wide business categories, and big economic volume. TikTok Pay was launched in January 2021, adding to the options of Alipay and WeChat Pay. Meituan has over 200 million daily active users, providing one-stop services for eating, drinking, and entertainment. Didi Taxi is a one-stop travel platform covering various businesses, changing the traditional way of taking taxis, and developing modern travel in the era of mobile Internet (Fig. 6—21).

6.5.3.8 Livable cities for humanity

The goal of digital urban governance is to create a livable city that is intelligent, green, and cultural. The metaverse, as the digital carrier of

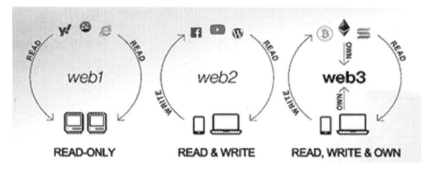

Figure 6–21 Web1.0, Web2.0, and Web3.0.

harmonious coexistence between man and nature, will be the entrance of future life where the material life, ecological environment, and humanistic spirit can be integrated.

In October 2021 Facebook officially changed its name into Meta, and aimed to build itself a Metaverse company in about 5 years. The Korea government created Metaverse alliance, and issued the "Basic Plan for Metaverse Seoul" to vigorously promote Korean industrial development. While in China in 2022, all large cities accelerated the introduction of policies to promote the integrated development of Metaverse technologies and applications.

6.5.4 Conclusion

Smart city development is in great need but still has a long way to go. According to a new report, the world's population is approaching 8 billion with more than 4 billion people in cities. In many large cities, "urban disease" has already been a serious obstacle to the economic and social development, as well as people's livelihood and environment improvement. The continuous integration and innovation of 4N4F will eventually give us some more intelligent "solutions" to develop a better city and have a better life!

6.6 Practice of four network four flows integration: an approach to digital Renaissance

4N4F integration is a scientific theory that summarizes the historical perspective of the laws of human social development and embodies considerable wisdom. It is also a practical methodology with a wide range of application scenarios in the current social development context. This chapter mainly expounds on the practical application of 4N4F integration theory in specific scenarios. Through these realizable projects, the practice of 4N4F will gradually guide the integration of the physical, humanity, and cyber worlds, ultimately leading to the great digital Renaissance of human society.

6.6.1 Four network four flows visualization: promotion to cyber and physical worlds integration

The integration of 4N4F is a sophisticated and grand scientific theory. However, the concepts involved, such as energy, transportation, information, humanities, are relatively abstract, and belong to different academic fields in traditional scientific disciplines, which increases the difficulty of communication and collaboration. To promote the wider application of 4N4F in practical projects, we hope that the theory of 4N4F integration can be understood by researchers, engineers, government officials, media personnel, and workers in related industries, and communication barriers caused by abstract content and cross-disciplinary communication can be avoided. Therefore the first step in putting the theory of 4N4F into practice is to create a "visualization" interface that can visualize abstract concepts and flatten cross-disciplinary knowledge and associate the rich cyber world it encompasses with our physical world in real life.

With the progress and maturity of digital technology, we have more ways to connect the real and virtual worlds. One of the widely spread concepts in recent years is the "metaverse." The metaverse is a hypothetical iteration of the Internet as a single, universal, and immersive virtual world that is facilitated by the use of virtual reality (VR), augmented reality (AR), and mixed reality (MR) headsets [45,46]. In colloquial usage, a "metaverse" is a network of 3D virtual worlds focused on social and economic connection [47−49]. The "metaverse" is not an independent new technology, but an integration of a large number of existing technologies, including 5G, cloud computing, artificial intelligence, VR, blockchain, digital currency, the IoT, human−computer interaction, and so on [50]. We will use these technology groups to build a visualized interface for 4N4F integration.

XR (extended reality) is the sum of VR, AR, and MR technologies, which are various virtual imaging technologies that have been used in reality in recent years. Among them, AR and MR are more advanced technologies that enable people to directly overlay and display virtual information such as graphics and text in the physical world scene through a smartphone, tablet, or other camera-equipped display, or a head-mounted device with transparent glasses, in such a way that it can be observed by the naked eye, as shown in Fig. 6−22.

MR technology emerged later than AR and is more advanced: AR typically presents noninteractive information, while MR presents information in scenes that are interactive and editable. For example, in a game called "Blaster," reported as the world's first sports game based on MR technology, developed by NetEase Games Innovation Lab in 2020, interactive virtual information such as health points, ammunition, virtual obstacles, and so on, can be perfectly overlaid onto real-world scenes using MR technology, and it achieves interactive gaming effects that are just as realistic as real props, shown in Fig. 6−23. Through the extended reality

Figure 6−22 Static augmented reality (AR) on mobile phone and dynamic mixed reality (MR) on wearable glasses.

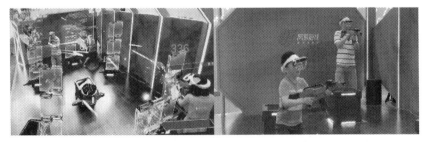

Figure 6−23 Scene from mixed reality (MR) glasses (left) and from third-person perspective (right).

Figure 6—28 Overlaying multiple periods of traffic info in visualized digital twin.

In the real world, there are no "ifs," but scientific research needs to constantly propose multiple hypotheses and verify them one by one to make progress. The digital twin can and does often exist before there is a physical entity. The use of a visualized digital twin in the creative phase allows the intended entity's entire life cycle to be modeled and simulated [53]. If complex comprehensive problems in the real world, such as environmental issues, can be repeatedly simulated and different results can be viewed in the laboratory, then scientists can make faster and more accurate scientific judgments.

With the increasing precision, richness of detail, and optimization of algorithms in digital twin technology, it is entirely possible to use a visualization digital twin system to complete the extrapolation of complex systems such as the environment. For example, when observing changes in carbon emissions in a community, we can import various factors such as the increase in population, vehicles, and industry, and the addition of green spaces and carbon storage conversion devices into the digital twin model in real-time. This allows for the intuitive observation of multiple potential outcomes in different settings, such as the effects of interventions on population, greenery, and economics. By comparing the various results produced by different factors, as shown in Fig. 6—29, government decision-makers can better develop policy guidance or intervention strategies.

Summary: The visualized 4N4F integration system, presented by technologies such as extended reality and digital twin, has brought about significant value and meaning by merging the previously abstract disciplines

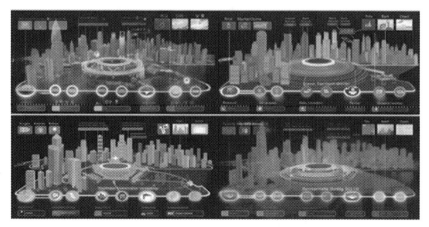

Figure 6–29 In digital twin, different policy interventions lead to different outcome.

of energy, transportation, and other fields into a tangible and visible form for the first time in the history of science:

Firstly, the visualized integration system breaks down the barrier between the cyber world and the physical world, promoting the integration of the two, and realizing a more effective scientific observation method, making it a valuable scientific practice.

Secondly, the visualized integration system is more easily understandable, allowing more ordinary learners to join the interdisciplinary integration research, providing a visible, operable, and discussable platform for the connection and communication between and within different disciplines, thus efficiently promoting the integration between interdisciplinary subjects.

Thirdly, based on the visualization of the integrated 4N4F system, the superimposition of multiple pieces of information can generate new information and knowledge, including data results that are difficult to obtain through traditional detection methods, such as the superimposition of multiple time periods and multiple possibilities. This has a significant promotional effect on scientific research or engineering advancement.

Fourthly, the visualization system provides a basic operating interface for the practice of the 4N4F integration systems and lays the foundation for further development. It is foreseeable that a value system that is computable, optimizable, and adjustable will be established based on the visualization, which will be discussed in detail in the next section.

6.6.2 Four network four flows value system: promotion to cyber and humanistic worlds integration

Based on the 4N4F visualization system, an evaluation mechanism for its integration results to achieve continuous optimization of the system will be further established.

In essence, it is hoped that the cross-border integration of energy, information, transportation, and other fields will evolve toward a more valuable direction. In the process of defining its "value," more human-oriented and humanistic factors should also be included in the evaluation criteria, such as more convenient, smarter, lower carbon, more orderly, and so on. In this section, a more humanistic evaluation system to guide the integration and progress of multidisciplinary fields and promote their transformation with the human world will be established.

In modern society, with the development of science and technology, there are increasingly more specialized fields within each discipline, which greatly enriches the diversity of systems while increasing the complexity of the conversion between multiple device systems within the same field. How to select the objects and methods for integration and transformation, and how to make the results of integration and transformation more meaningful and valuable are the first questions to be answered in the practice of 4N4F integration.

In the field of energy, for example, in modern cities, civilian energy is mainly divided into three categories: electricity, gas, and heat, each of which is further divided into more subsystems. For example, electricity can be divided into DC systems and AC systems; and DC systems can further include photovoltaic panels, lithium batteries, and other entities. In daily production and life, there are many processes of conversion, interaction, and coordination among different energy sources. In most cases, users only consider the loss rate and energy price as the main evaluation criteria. However, in the value concept of the 4N4F integration, we hope to include many evaluation factors such as safety, user convenience, balance of source-grid-load-storage, user friendliness, environmental friendliness, and the possibility of sustainable development, to guide the establishment of smart energy systems that meet the value standards of the humanity network. We use a formula to illustrate the evaluation mechanism as follows:

$$V = \alpha_1 N + \alpha_2 S + \alpha_3 C - \alpha_4 LR - \alpha_5 FR - \alpha_6 M \pm \alpha_7 E$$

This is a guideline-based open formula designed to provide a methodological framework for value evaluation. Users should adjust the coefficients and parameters based on specific project practices and circumstances. In the formula, N represents necessity (such as peak-valley fluctuation and shortage of converted energy); S represents safety; C represents convenience (such as the usability, versatility, or adaptability of converted energy); LR represents loss rate of the energy conversion, FR represents the failure rate or other safety factors during the conversion process; M represents additional management costs brought by the conversion process; E represents the evaluation of pollution, carbon emissions, and overall environment before and after the conversion (the positive or negative sign represents whether the impact on the environment is more friendly or adverse). $\alpha 1$-$\alpha 7$ represent their respective weighting coefficients. The comprehensive score V calculated is the value evaluation result of the current energy conversion. If V is greater than zero, it means that the current energy conversion has produced beneficial value and should be promoted. Otherwise, it has brought harmful effects to the whole and should be further optimized. The equation is shown in Fig. 6−30.

A more complex evaluation mechanism than a single-domain system is the evaluation mechanism for cross-disciplinary projects. Moreover, because it involves the barrier of setting standards across disciplines, unified evaluation criteria with a common value system have yet to be established in the category of cross-disciplinary projects. From the perspective of comprehensively guiding the practice of four-network and four-flow integration and overall layout of the humanistic network and value flow, filling the gap in the evaluation mechanism across disciplines is particularly important.

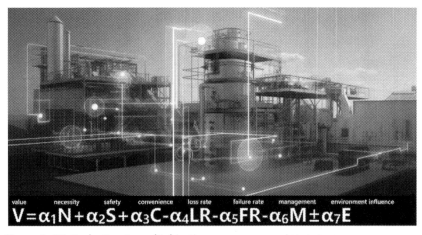

$$V=\alpha_1 N+\alpha_2 S+\alpha_3 C-\alpha_4 LR-\alpha_5 FR-\alpha_6 M\pm\alpha_7 E$$

Figure 6−30 Evaluation in multiobjective energy conversion system.

Taking intelligent connected vehicles (ICV) as an example, it refers to the organic combination of the IoVs and smart cars, which ultimately replaces humans as operators of the new generation of vehicles. ICVs are equipped with advanced on-board sensors, controllers, actuators, and other devices, integrating modern communication and network technologies to achieve intelligent information exchange, and sharing between vehicles, humans, roads, and back-end systems, with the characteristics of safety, comfort, energy conservation, and high efficiency. Its emergence is the result of large-scale cross-domain integration. This complex system includes humanistic requirements for liberating human hands and creating convenient travel, technological challenges for breakthroughs in key areas such as energy and transportation efficiency, and the complex integration among them and other disciplines. In this process, we also hope to establish a formula for evaluating its value, to comprehensively evaluate its pros and cons, and to provide the basis for system optimization with calculations and data.

$$V = \alpha_E V_E + \alpha_T V_T + \alpha_S \Delta S + \alpha_C \Delta C + \alpha_{El} \Delta E - \alpha_p \Delta P - \alpha_m \Delta M$$

Like evaluation formulas within a single domain, this is also an open formula for guideline-based evaluation. In the formula, V_E represents the value evaluation score in the energy field, which should at least include complex conversions such as the national grid (AC), charging pile (DC), lithium battery (DC), and V2X system; V_T represents the value evaluation score in the transportation field, which should at least include infrastructure investment, transportation efficiency improvement, vehicle-road coordination enhancement, and improvement of human−vehicle travel routes; ΔS, ΔC, ΔE represent the increase in safety, convenience, and environmental friendliness before and after the promotion of ICV, respectively, while ΔP represents the increase in the cost of the entire system, and ΔM represents the increase in urban management costs of the entire system. The corresponding α coefficients still represent the weight of each parameter. Like the previous formula, the comprehensive score V calculated based on these factors is the value evaluation result of this cross-disciplinary integration. If V is greater than zero, it means that it has produced beneficial value and should be promoted; otherwise, if it has a harmful impact on the overall system, it should be further optimized.

From the expression of the formula, it is not difficult to see that in the exploration of cross-domain integration of ICV, we have a high probability of obtaining a positive value judgment. Moreover, if we continue to optimize the system toward more safety, convenience, intelligence, and low-carbonization, adhere to people-oriented, and further optimize the technology, we can continue to promote greater value brought by the ICV industry (Fig. 6–31).

It is foreseeable that as systems become increasingly complex and the scope of cross-disciplinary integration expands, the number of physical quantities and value points that need to be considered will also increase. If these complex systems are simply compared based on data, it will be difficult to establish a unified measuring standard within different discipline. However, if everything is guided by "humanistic values" and oriented toward the symbiotic and harmonious development of humans and nature, we can establish a more concise and universal evaluation principle as our core value. From the perspective of the classical third law of thermodynamics, this is the principle of "minimizing the overall entropy increase caused by completing the same energy efficiency" [54].

$$V_{MAX} \rightarrow \Delta E_{MIN}$$

ΔE represents the system entropy increase. It is not only about saving at the technical level, such as reducing losses, improving efficiency, and simplifying systems; but also about restraint at the humanistic value level, such as reducing unnecessary complicated operations, satisfying more

Figure 6–31 Evaluation in intelligent connected vehicles (ICV) cross-disciplinary conversion.

human needs at once, and promoting low-carbon sustainable development. In other words, the more a system meets requirements such as low loss, high efficiency, simple operation, environmental friendliness, multifunctional capabilities, and sustainable development while achieving the same energy efficiency, the more it should be advocated; otherwise, it should be adjusted or optimized. Guided by such a core value, even in the face of more complex integrations in the future, we can always find the right direction for progress. Fig. 6–32 illustrates the universal value principle from the entropy increase perspective.

Summary: Built upon a visualized interface, a further computable and optimizable humanistic value evaluation system will be established to promote the development of disciplines and cross-disciplinary fields toward more ordered and optimized directions. Its significance is:

Firstly, the establishment of a value evaluation system is an important part of the value flow in the 4N4F integration. Under the guidance of more advanced, intelligent development perspectives, and humanistic care, we will better promote the integration of social development into the humanistic world.

Secondly, the value flow system provides evaluation standards of good and bad, which will help in the practice of 4N4F projects with multiple disciplines and parameters, providing a way to screen and optimize for the possible different forms of integration, ensuring sustainable and optimized project development.

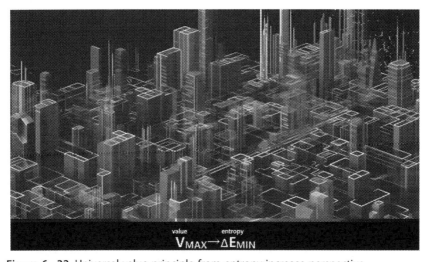

Figure 6–32 Universal value principle from entropy increase perspective.

Thirdly, the value evaluation system provides a basis for directed evolution, which will guide each integration field with certain values to obtain the correct methodology for optimization and evolution from the beginning of its formation, with greater opportunities to reach higher heights in its future development and practice.

Fourthly, only under the guidance of a value system that combines technological innovation and humanistic care, will the complexity of the 4N4F integration system transform into "wisdom." The 4N4F practice will promote the establishment of multiple intelligent systems such as "smart energy," "smart transportation," "smart information," and ultimately form a greater "smart city," which we will describe in detail in the next section.

6.6.3 Four network four flows—based smart city: a comprehensive integration of physical, humanistic, and cyber worlds

A smart city is a more advanced form of a city that integrates the constituent systems and services of the city using various information technologies or innovative ideas to enhance the efficiency of resource utilization, optimize urban management and services, and improve the quality of life of citizens. Smart cities fully utilize new information technologies in various industries of the city and are the advanced form of urban informatization based on innovation in the knowledge society, realizing the deep integration of informatization, industrialization, and urbanization. They help alleviate "urban diseases," improve the quality of urbanization, achieve refined and dynamic management, and enhance the effectiveness of urban management and the sustainable innovation ecology to improve the quality of life of citizens [55].

The smart city concept integrates humanity with information and communication technology (ICT), and various physical devices connected to the IoT network to optimize the efficiency of city operations and services and connect to citizens [56]. Smart city is the comprehensive integration of the physical world, cyber world, and humanistic world. It is both a vision of a better future for humanity and an inevitable outcome of the rapid progress in the IoT, ICT, mobile internet, digital technology, and humanistic openness, all of which have been guiding the development of cities, as shown in Fig. 6—33. The core and goal of smart cities are human-centered, harmonious coexistence between humans and nature, and sustainable development; its foundation is the integration of

Figure 6-33 Smart city concept, with physical, humanistic, and cyber worlds integrated.

the physical, cyber, and humanity worlds, which is 4N4F convergence. The reasons for this are as follows:

The concept of a smart city requires a significant improvement in the operational efficiency and quality of integrated support services based on existing urban forms. To achieve this goal, it is necessary to realize cross-domain collaboration. With the integration of multiple fields such as energy, transportation, and information, the optimization of energy utilization can be achieved through an intelligent energy management system, traffic congestion can be alleviated, and road transportation efficiency can be improved. At the same time, the high-speed and stable network infrastructure can be used to achieve interconnection of information in various urban fields, thereby promoting resource optimization and cross-domain collaborative development. Such close integration helps improve urban management efficiency and enhance overall urban efficiency, as shown in Fig. 6-34.

Smart cities are built on a value foundation that puts people first, with the fundamental goal of enhancing people's happiness and well-being. The 4N4F initiative advocates for the establishment of a humanistic network and the implementation of human-centered urban governance, focusing on individual needs and dignity, fully unleashing people's

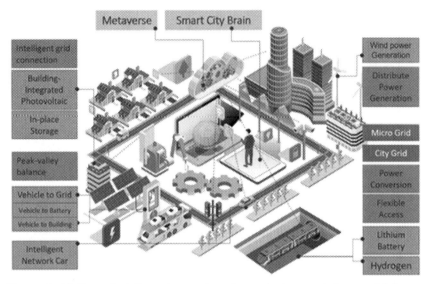

Figure 6–34 Cross-disciplinary integration of energy, transportation, and information enhances the efficiency and quality of the entire city.

subjective initiative, emphasizing social fairness and justice, providing all-around care and support, and enriching people's cultural life.

By focusing on the value and needs of everyone, ensuring social fairness and justice, encouraging diverse cultural activities, and establishing a sound social security system, we create a more harmonious and beautiful living environment for people, thereby improving people's sense of happiness, achievement, and satisfaction.

Smart cities must be sustainable. In the theory of 4N4F, the value stream guided by lower carbon emissions and greater order advocates for the intensive application of all societal resources, which can help improve the efficiency of resource and energy use, promote wider adoption of smart energy systems, stimulate green innovation, reduce environmental pollution, enhance public environmental awareness, and drive global cooperation. The value stream also encourages the improvement of resource utilization efficiency in the production, consumption, and service sectors, reducing waste, while stimulating green technology innovation and advocating for green lifestyles. By reducing reliance on natural resources, protecting the ecological environment, optimizing industrial structure, and strengthening international cooperation, economic development, and environmental protection can be coordinated to achieve sustainable socio-economic development, smart cities will get smarter [57].

The evolution direction of smart cities is the 4N4F integration. Therefore the process of integrating is an effective means of promoting the formation of smart cities. This integration process involves various aspects of socio-economic development, and this section only selects several representative directions as examples:

Technological progress and productivity development are necessary foundational conditions for building smart cities. They play a decisive role in promoting urban transformation and upgrading, improving urban management, optimizing the urban environment, and enhancing the quality of life for citizens.

The primary task of the 4N4F practice is to promote cross-disciplinary integration. This innovative integration provides opportunities for technological progress, breaks through disciplinary limitations, improves research efficiency, and helps solve more complex problems. At the same time, during the process of cross-disciplinary integration, cooperation between top experts in different fields breaks down existing knowledge boundaries, creates new scientific research results, and further disseminates new knowledge through the power of people working together. Therefore the process of 4N4F integration will greatly promote technological progress and the development of productivity and create necessary foundational conditions for the arrival of smart cities.

City infrastructure refers to the sum of engineering and social infrastructure necessary for the survival and development of a city. It includes all kinds of facilities built in the city for the smooth conduct of various economic and other social activities. The existing infrastructure in cities such as roads, parking lots, energy networks, and information base stations are the basic hardware for expressing city functions. Today, with the highly developed digital technology, the infrastructure built in the previous era generally exhibits a lack of software support. However, by creating a visualized interface for the city 4N4F integration, a "digital infrastructure" that matches the hardware infrastructure can be created as software support. It is a software-based online infrastructure that facilitates interactions and transactions between users [58] and will greatly enhance the efficiency of the existing city infrastructure and create a material basis for achieving a smart city. Acted in digital twins, this digital infrastructure can be often proposed in the form of interactive platforms to capture and display real-time 3D and 4D spatial data in order to model urban environments and the data feeds within them [59].

Using transportation infrastructure as an example, if all effective public parking spaces in the city are unified and recorded in the system, it is possible to integrate roadside parking spaces, parking garages underground parking, office parking, community public parking, and some privately owned parking, with the consent of the owners, into a unified parking resource library. With the help of artificial intelligence algorithms to allocate parking spaces and reserve parking spaces for all citizens, and by conveying information to drivers through various navigation interfaces, the maximum utilization of all parking resources in the city can be achieved, thus freeing up idle spaces in the city, improving overall efficiency, and effectively solving the problem of parking difficulty.

Taking the information infrastructure as an example, following the same idea, all service information related to city services, including living facilities, logistics, healthcare, education, transportation, police, and civil affairs, can be collected on a unified city visualization information platform, forming a "smart city brain" that provides comprehensive smart services for the entire city. This new type of digital infrastructure brings benefits to both citizens and city managers: citizens can enjoy more intuitive and convenient city services, while city managers can use it to allocate public resources, make scientific decisions, and improve governance efficiency [60], as shown in Fig. 6—35.

"Suitable for living and business" are the two widely accepted standards for measuring smart cities [61]. One of the indicators of "suitable for business" is the completion of industrial upgrading, ecological aggregation, and integration of industry and city. Through cross-disciplinary project cooperation in the 4N4F integration, this process can be promoted.

Firstly, multidisciplinary cooperation will promote urban industrial upgrading. Multidisciplinary cooperation helps to break the boundaries of traditional industries, and introduce new technologies and methods, so that

Figure 6—35 4N4F visualization helping to enhance the city infrastructure efficiency: smart parking navigation (left) and smart city information brain (right).

enterprises can better meet market demand. This cooperation can promote enterprises to carry out technological innovation, product upgrading, and service optimization, thus realizing the transformation and upgrading of the industry and improving its added value and competitiveness.

Secondly, multidisciplinary cooperation will promote ecological aggregation. Cooperation can drive the participation of enterprises, research institutions, and government departments in various fields, forming a collaborative development in various aspects of the industrial chain, such as research and development, production, and sales. This collaborative development is conducive to the formation of an industrial ecosystem, attracting more high-quality enterprises and talents to gather, further promoting the development and growth of the industry.

Thirdly, multidisciplinary collaboration can promote the integration of industry and city. Collaborative projects across different disciplines can promote the close integration of industry and urban infrastructure, public services, and talent cultivation. Through the interaction between industry and city, citizens' needs can be better met, and the attractiveness and vitality of the city can be enhanced. At the same time, the integration of industry and city helps to achieve efficient use of resources, improve the level of sustainable urban development, and further promote the construction of smart cities, as shown in Fig. 6−36.

"Suitable for living" is another benchmark for measuring smart cities, including the convenience of people's lives, the improvement of their happiness, and the sustainable development of the urban environment [62].

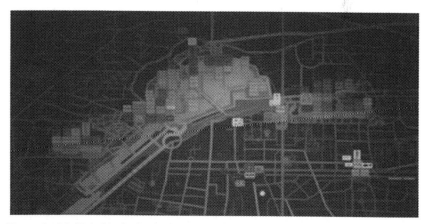

Figure 6−36 "Suitable for business" city with industry ecological clustering.

Through the guidance of the "Human Network" and "Value Flow" in the four networks and four streams, this goal can be achieved, as shown in Fig. 6–37.

Firstly, humanity values will improve the convenience of city life. Under the guidance of human values, urban planning and management will pay more attention to the needs and comfort of residents. For example, by providing convenient public transportation, improving community service facilities, and intelligent living services, a convenient and comfortable living environment can be created for residents. Human care can also promote the optimization and diversification of public spaces, providing more leisure and cultural activity spaces for residents.

Secondly, humanity values will enhance people's happiness and sense of acquisition. Human values advocate for the all-round development of people, caring about the physical and mental health and quality of life of residents. By improving the level of public services such as education, medical care, and social security, residents can enjoy higher life security. In addition, human care emphasizes inclusiveness and diversity, encourages respect and understanding of different cultures, religions, and lifestyles, and jointly creates a harmonious and inclusive social atmosphere, thereby improving people's sense of happiness.

Thirdly, humanity values will achieve sustainable development of cities and society. Urban development guided by human values emphasizes seeking a balance between economic growth and environmental protection and pursues sustainable development. For example, promoting green travel, green buildings, energy conservation and emission reduction measures, reducing pressure on the environment, and leaving room for

Figure 6–37 "Suitable for living" city with convenience, happiness, and sustainability.

sustainable development for future generations. In addition, under the guidance of the value flow, the cross-disciplinary integration direction of urban projects will also move toward resource-intensive applications and long-term sustainable development.

The goal of 4N4F integration practice is highly consistent with the vision of building a smart city. The 4N4F integration is both a core content of smart cities and an important process for promoting the formation of smart cities. The relationship and role between them are as follows:

Firstly, the 4N4F integration and the vision of smart cities are both to achieve more advanced human cities and social forms by promoting the comprehensive integration of physical world, cyber world, and humanity world.

Secondly, the core content of a smart city is the improvement of urban efficiency, the happiness of the people, and the sustainable development of humans and nature. 4N4F integration can become an important core path to promote the construction of smart cities by promoting urban productivity development, improving infrastructure efficiency, and realizing the "suitable for business and living" goal.

Thirdly, a smart city is the carrier of human wisdom society and the sum of the most advanced science and technology and human achievements. It is based on the results of the Fourth Industrial Revolution, namely the digital technology revolution, and marks another takeoff of human civilization. 4N4F integration plays an important role in the process of digital Renaissance and the formation of smart cities, which will be further discussed in the next section.

6.6.4 Four network four flows practice: an approach to great digital Renaissance

Throughout human history, great revivals or major changes have often been driven by a significant breakthrough with decisive significance, which gradually evolves in collaboration with other fields, such as the Renaissance, which was awakened by the humanist spirit, and led to the prosperity of art, science, politics, and economics, bringing humanity from the Middle Ages into a civilized society [63]. The Industrial Revolution, on the other hand, was driven by significant technological breakthroughs such as the steam engine, which comprehensively led to advancements in industry, agriculture, trade, and urbanization, allowing humans to make the second major leap and enter modern society [64]. The era we now face is a historic moment when digital and internet technology has formed explosive growth, and the cyber world is about to fully integrate with our existing world.

In the previous section of this chapter, we reviewed how the 4N4F integration promotes the convergence of the digital world: visualization of digital technology based on interdisciplinary fields such as transportation, energy, and information builds a basic operating interface for the integration of the cyber and physical worlds; optimization of data and value guidance during the integration process constructs the value core for the integration of the cyber and humanity worlds; in the process of building smart cities, more grand projects integrating across boundaries provide application scenarios for the comprehensive integration of the physical, cyber, and humanity worlds. Through extensive integration practices of 4N4F, human society will embark on a new path of a great "Digital Renaissance." The main significance is as follows:

6.6.4.1 Practice of four network four flows integration is a Renaissance from zero to one

The first aspect of "Renaissance" is the breakthrough from nothing to something. Through creative cross-domain integration of various technological innovations, such as the IoT, artificial intelligence, MR, 5G technology, and humanized value guidance, human society has achieved the integration of physical world, cyber world, and humanity world for the first time. This is a historic breakthrough from zero to one, a process of qualitative change brought about by quantitative change. It not only demonstrates that "cross-domain integration" is a highly valuable methodology, but also realizes a highly innovative social model, and deduces an unprecedented model of human "smart cities."

6.6.4.2 Practice of four network four flows integration is a Renaissance from fragment to holism

The second aspect of "Renaissance" is that it promotes the evolution of the knowledge system of human society from independent layouts to a comprehensive architecture of integration. In the practical projects of 4N4F integration, not only has interdisciplinary communication been promoted in the research level, but also the specific technical interfaces between multiple systems have been further refined throughout the entire process of project implementation through practical applications. The law of scientific development tells us that the integration of disciplines is not just an increase in quantity; more importantly, the result of integration is that the gaps between various fields are filled, and a large amount of new knowledge, new technologies, and new sciences will emerge, which may even lead to an explosive growth of new human technologies, bringing about a major leap in the scientific and technological system.

6.6.4.3 Practice of four network four flows integration is a Renaissance from data to value

The third aspect of "Renaissance" is the progress from general data to value-oriented humanistic guidance. We believe that the values are more ordered data [65], and that wisdom is a combination under specific value guidance. The progress of our social economy should not only focus on changes in numerical values, but also on whether it has positive humanistic value guidance, such as being more low-carbon, environmentally friendly, orderly, convenient, livable, and business-friendly, fair and just, and improving people's happiness and sense of achievement, among other things. We propose the principle of minimizing the entropy increase in the entire system and expressing it in formulaic form. This principle unifies multiple data that originally belonged to different evaluation criteria, such as improving system efficiency, reducing total consumption, avoiding environmental damage, and enhancing humane governance, into a comprehensive value evaluation system under the guidance of the humanistic network and value flow. This system guides and realizes the sustainable development of the entire social economy.

6.6.4.4 Practice of four network four flows integration is a Renaissance of questioning spirit

The fourth aspect of the "Renaissance" is the re-emergence of human questioning in the exploration of the world, and from this questioning, the answer to the next generation of urban development and technological revolution has been found [66]. In modern society, since the Industrial Revolution, the principle of division of labor, independent disciplines, and clear management has greatly improved the overall efficiency and quality of society. However, under the rapidly emerging new technology represented by digital technology, the necessity and urgency of "integration" between industries and fields are becoming increasingly prominent, and the traditional industry division and urban operation mechanisms are being challenged.

Under the guidance of the "people-oriented" spirit, we boldly proposed and gradually demonstrated hypotheses for cross-disciplinary and cross-field cooperation, forming the theory and practice of "4N4F integration" to respond to this historic challenge. Through digital presentation technology, we have completed the construction of the visualized 4N4F interface, lowering the threshold for understanding the theory itself, increasing its popularity and applicability. Based on this, we have formed an evaluative system that is computable, discussable, and can be used to guide judgments, to continuously optimize the entire system toward a more humanistic direction, thus providing a value-driven core. The 4N4F integration system, with its visual operating interface and core value flow guidance, will play an important role in the construction of "smart cities" by humans, and ultimately achieve a new and efficient, low-carbon, livable, and business-friendly urban and social patterns. This is bold thinking and a great exploration of the next generation of urban and social models from a human perspective.

6.6.5 Conclusion

Looking back at the previous content, the theory of the 4N4F integration comes from the observation and judgment of human social development and natural scientific evolution trends over thousands of years. It is a logical speculation with a considerable historical development perspective. The practice of 4N4F integration is a process of gradually verifying and implementing the theory, and the ultimate result is to achieve tremendous progress in human technology, productivity, and social forms. It aims to build a smart city where the physical, cyber, and humanity worlds are

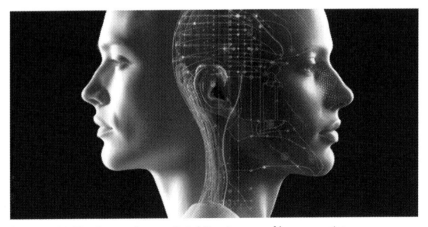

Figure 6–38 The future of great digital Renaissance of human society.

fully integrated, and to realize a human-centered, harmonious coexistence between humans and nature for sustainable development in human society, as shown in Fig. 6—38. It comes from history and will lead us to open the future of a great digital Renaissance.

References

[1] China Southern Power Grid Co., Ltd. (CSG), Digital grid promotes the construction of a new power system with new energy as the main body white paper, 2021.

[2] S.S. Reka, T. Dragicevic, Future effectual role of energy delivery: a comprehensive review of internet of things and smart grid, Renew. Sustain. Energy Rev. 91 (2018) 90—108.

[3] K. Qian, T. Lv, Y. Yuan, Integrated energy system planning optimization method and case analysis based on multiple factors and a three-level process, Sustainability 13 (2021) 7425.

[4] B. Bilgin, A. Emadi, Electric motors in electrified transportation: a step toward achieving a sustainable and highly efficient transportation system, IEEE Power Electron. Mag. 1.2 (2014) 10—17.

[5] A. Albatayneh, et al., Comparison of the overall energy efficiency for internal combustion engine vehicles and electric vehicles, Rigas Tehniskas Universitates Zinatniskie Raksti 24.1 (2020) 669—680.

[6] J. Contreras-Castillo, S. Zeadally, J.A. Guerrero-Ibañez, Internet of vehicles: architecture, protocols, and security, IEEE Internet Things J. 5.5 (2017) 3701—3709.

[7] M. Maurer, et al., Autonomous Driving: Technical, Legal and Social Aspects, Springer Nature, 2016.

[8] P.A. Hancock, Imposing limits on autonomous systems, Ergonomics 60.2 (2017) 284—291.

[9] M.R. Endsley, Situation awareness in future autonomous vehicles: beware of the unexpected, Congress of the International Ergonomics Association, Springer, Cham, 2018.

[10] P. Reilhac, et al., User experience with increasing levels of vehicle automation: overview of the challenges and opportunities as vehicles progress from partial to high automation, Automot. User Interfaces (2017) 457—482.

[11] C. Roth, *Design of the in-vehicle experience.* No. EPR2022012. SAE Technical Paper, 2022.

[12] D. Gorecky, et al., Human-machine-interaction in the Industry 4.0 era, in: 2014 12th IEEE International Conference on Industrial Informatics (INDIN), IEEE, 2014.

[13] G. Johannsen, Human-machine interaction, Control. Systems Robot. Autom. 21 (2009) 132—162.

[14] A. Bauer, D. Wollherr, M. Buss, Human—robot collaboration: a survey, Int. J. Humanoid Robot. 05 (01) (2008) 47—66. Available from: https://doi.org/10.1142/s0219843608001303.

[15] CDI Products, How cobots are powering smart manufacturing, 2022. https://www.cdiproducts.com/blog/how-cobots-are-powering-smart-manufacturing.

[16] J.E. Colgate, W. Wannasuphoprasit, M.A. Peshkin, Cobots: robots for collaboration with human operators, in: Proceedings of the 1996 ASME International Mechanical Engineering Congress and Exposition, 1996.

[17] Collaborative Robot Market. (n.d.). MarketsandMarkets. Link: https://www.marketsandmarkets.com/Market-Reports/collaborative-robot-market-194541294.html.

[18] D. Liu, J. Cao, Determinants of collaborative robots innovation adoption in small and medium-sized enterprises: an empirical study in China, Appl. Sci. 12 (19) (2022) 10085. Available from: https://doi.org/10.3390/app121910085.

[19] R. Müller, M. Vette, O. Mailahn, Process-oriented task assignment for assembly processes with human-robot interaction, Procedia CIRP 44 (2016) 210–215. Available from: https://doi.org/10.1016/j.procir.2016.02.080.
[20] S. Nahavandi, Industry 5.0—a human-centric solution, Sustainability 11 (16) (2019) 4371. Available from: https://doi.org/10.3390/su11164371.
[21] M. Schnell, M. Holm, Challenges for manufacturing SMEs in the introduction of collaborative robots, Adv. Transdisciplinary Eng. (2022). Available from: https://doi.org/10.3233/atde220137.
[22] Automate, The role of collaborative robots in Industry 5.0. (n.d.). https://www.automate.org/blogs/the-role-of-collaborative-robots-in-industry-5-0.
[23] W. Gu, et al., Modeling, planning and optimal energy management of combined cooling, heating and power microgrid: a review, Int. J. Electr. Power Energy Syst. 54 (2014) 26–37.
[24] V. Marinakis, et al., From big data to smart energy services: an application for intelligent energy management, Future Gener. Comput. Syst. 110 (2020) 572–586.
[25] T. Ahmad, et al., Artificial intelligence in sustainable energy industry: status quo, challenges and opportunities, J. Clean. Prod. 289 (2021) 125834.
[26] C.-C. Chan, Guiding the new era of electrified transportation with 4-networks 4-flows integration theory and practice, J. Glob. Tour. Res. (2021).
[27] Q. Yue, L. Shuai, L. Hai, L. Enbo, G. Wei, Z. Wennan, Flexibility of integrated energy system: basic connotation, mathematical model and research framework, Autom. Electr. Power Syst. 46 (17) (2022) 16–43.
[28] H. Huachun, W. Shengjun, W. Chenggen, Development of integrated energy online simulation system based on joint calculation and analysis of multi-energy flows, Power Demand. Side Manag. 23 (04) (2021) 33–38.
[29] W. Mengxue, Z. Haoran, T. Hang, C. Jian, W. Qiuwei, Review of typical simulation and planning platforms for integrated energy system, Power Syst. Technol. 44 (12) (2020) 4702–4712.
[30] S. Hongbin, G. Qinglai, W. Wenchuan, W. Bin, X. Tian, Z. Boming, Integrated energy management system with multi-energy flow for Energy Internet: design and application, IEEE Electrific. Mag. 43 (12) (2019) 122–128.
[31] S. Wang, Intelligent city research methods for urbanization, Planners 28 (04) (2012) 19–23.
[32] Z. Sun, F. Zhen, Intelligent city development and planning practice review, Planners 29 (02) (2013) 32–36.
[33] S. Myeong, J. Park, M. Lee, Research models and methodologies on the smart city: a systematic literature review, Sustainability 14 (3) (2022).
[34] UNESCO. Man and Biosphere Programme (MAB), 1971.
[35] Louzhishouju, Quaternary Sector of Economy, China Citic Press Group, Beijing, 2022.
[36] S. Acar, Matthew, Pygmalion, and Founder Effects, Academic Press, San Diego, 2011, pp. 75–81.
[37] M. Lewis, The City in History, Shanghai Sanlian Bookstore, Shanghai, 2018.
[38] M. Csikszentmihalyi, Flow: The Psychology of Optimal Experience, Harper & Row, 1990.
[39] A. Pentland, Social Physics: How Good Ideas Spread-the Lessons from a New Science, The Penguin Press, NY, 2014, p. 320.
[40] Technavio Smart City Market by Application and Geography — Forecast and Analysis 2021–2025. Technavio Co. Ltd, 2022. Available from: https://www.technavio.com/report/smart-cities-market-industry-analysis.
[41] K. Kloeckl, O. Senn, C. Ratti, Enabling the real-time city: LIVE Singapore! J. Urban. Technol. 19 (2) (2012) 89–112.

[42] M. Angelidou, Four European smart city strategies, Int. J. Soc. Sci. Stud. (2016) 18—30.

[43] EPRI, Power Delivery System and Electricity Markets of the Future, 1009102, EPRI, Palo Alto, CA, USA, 2003.

[44] E.A. Richardson, R. Mitchell, Green cities and health: a question of scale? J. Epilepsy 66 (2) (2012) 160—165.

[45] M. O'Brian, K. Chan, EXPLAINER: what is the metaverse and how will it work? ABC N. (2021).

[46] C. Newton, Mark Zuckerberg is betting Facebook's future on the metaverse, Verge (2021).

[47] G.D. Ritterbusch, M.R. Teichmann, Defining the metaverse: a systematic literature review, IEEE Access (2023). Available from: 10.1109/ACCESS.2023.3241809.

[48] A. Robertson, (2021). *What is the metaverse, and do I have to care?* Verge.

[49] P.A. Clark, What is the metaverse and why should I care? Time (2021).

[50] Y. Shen, China Xinhua News Network, http://www.news.cn/mrdx/2021-11/21/c_1310323484.htm, 2021.

[51] Y. Cai, Sensor data and information fusion to construct digital-twins virtual machine tools for cyber-physical manufacturing, Procedia Manuf. (2017).

[52] Wikipedia, Digital twin item, 2023, https://en.wikipedia.org/wiki/Digital_twin.

[53] M. Grieves, Can the digital twin transform manufacturing. World Economic Forum of Emerging Technologies, 2015.

[54] N. Georgescu-Roegen, The Entropy Law and the Economic Process, *Harvard University Press*, 1971.

[55] Wikipedia, "Smart city" item, 2023, https://zh.wikipedia.org/zh-my/smart%20city.

[56] M. Peris-Ortiz, D.R. Bennett, Yábar, D. Pérez-Bustamante, Sustainable Smart Cities: Creating Spaces for Technological, Social and Business Development, *Springer*, 2020.

[57] E.L. Glaeser, C.R. Berry, Why Are Smart Places Getting Smarter? *Harvard University*, 2018.

[58] Wikipedia, "Digital infrastructure" item, 2023, https://en.wikipedia.org/wiki/Digital_platform_(infrastructure).

[59] NSW, NSW digital win, Digital, 2020.

[60] Alibaba, The city brain: Practice of large-scale artificial intelligence in the real world, IET Smart Cities, 2019.

[61] IBM, Smart cities in China, white paper, 2020.

[62] McKinsey, "Smart city technology for a more liveable future." http://www.mckinsey.com, 2019.

[63] J. Monfasani, Renaissance Humanism, from the Middle Ages to Modern Times, *Taylor & Francis*, 2016.

[64] J. Horn, L. Rosenband, M. Smith, Reconceptualizing the Industrial Revolution, *MIT Press*, 2010.

[65] Neil Hampson, Creating Value from Data, PwC (2019). Available from: https://www.strategyand.pwc.com/gx/en/insights/2019/creating-value-from-data.html.

[66] J. Brotton, Science and philosophy, The Renaissance: A Very Short Introduction, *Oxford University Press*, 2006.

Index

Note: Page numbers followed by "*f*" and "*t*" refer to figures and tables, respectively.

A

Accessibility, 159
ADT. *See* Airframe Digital Twin (ADT)
Active management, 143
Advanced strategies, 187
Advanced technology, 231–232
AHP. *See* Analytical Hierarchy Process
    (AHP)
AI. *See* Artificial intelligence (AI)
Air pollution, 157
Air quality monitoring networks, 261
Airframe Digital Twin (ADT), 177–178
Analytical Hierarchy Process (AHP),
    172–173
AR. *See* Augmented reality (AR)
Architecture and organizational
    implementation, 10–11
Artificial carbon conversion, 13–14
Artificial intelligence (AI), 1–3, 27, 44, 47,
    58–59, 71, 75–77, 88–89, 89*f*,
    95–96, 137, 164, 171, 220–221,
    231–232, 235–236, 252–253
  challenges of artificial intelligence in
      renaissance, 52–56, 54*f*
  cloud platform, 80
  energy security from artificial
      intelligence–driven energy, 1–11, 7*f*
  in energy systems, 3–7, 5*f*
    decarbonizing power sector for full-
        system decarbonisation, 5–6
    digitalization, 4–5
    future power system, 6–7
  solutions, 10
  technology, 80, 185–186
    contemporary development and
        application of, 76–77
    integration of knowledge and data
        intelligence, 77*f*
Artificial neural network, 49–52
Augmented reality (AR), 78, 151–152,
    263

Automatic decision-making system, 199
Automatic driving system, 223–225
Automatic theorem-proving system, 1–2
Automatic tolling system, 260
Automation process, 233
Automation system, 237
Automobile popularization, 255
Automobile technologies, 220–221
Autonomous decision-making, 189
Autonomous driving, 214, 222–225
Autonomous system behaviour, 173–174

B

Barriers, 240
Batteries, 215
Beginning of Life (BOL), 171, 190
Behavior-driven smart vehicle, 212–228
  development of smart vehicles, 220–228
  future trends of electric vehicles,
      215–220
Big data, 27, 55–56, 75, 137, 143,
    173–174, 179–180, 261
  analytics, 26
  mining, 206–208
  platform, 258
  system, 95–96
Biofuels, 84–88
Black box models, 155–156, 182–183
Blockchain technology, 62, 69, 75–76, 81,
    149, 266–267
Bluetooth for vehicle-road network,
    221–222
BOL. *See* Beginning of Life (BOL)
Brain-machine interface, 114–115
Bus technology, 219–220
Business model, 105

C

CAN. *See* Controller area network
    (CAN)

Carbon, 11
    balance energy, 99
    cycle, 11, 16−18
    emissions, 159, 271−272
    material cycle, 14
    neutral energy system, 133, 134f
    neutralization, 84
    peaking, 11−12, 83−99, 85t, 138,
        204−205
    removal, 13
    sequestration, 13
    sink, 13−14
    storage conversion devices, 269
    technologies, 7−8
Carbon dioxide emissions, 5−6, 13, 83
Carbon neutrality, 13, 83−99, 204−205
    carbon neutralization schedule of count,
        85t
    energy ecosystem, 11−18, 17f
        closed carbon cycle of energy
            ecosystem, 17f
        coupling between energy flow and
            material flow, 18f
        development stages of energy
            applications, 15t
        roadmap, 15−16, 16f
    natural and energy ecosystem, 11−14
CCUS. See CO₂ capture, use and storage
    (CCUS)
Charging methods, 128
Charging pile, 273
Charging power, 128
Chips, 218−219
CHP. See Combined heat and power
    system (CHP)
CHPF. See Combined heat, power, and
    fuel (CHPF)
CIMI. See Cities in motion index
    (CIMI)
Cities in motion index (CIMI), 32
City infrastructure, 279
City-wide digital intelligent system, 83
Climate ambition, 84−88
    challenges faced by energy revolution,
        90−91, 93f
    fourth industrial revolution, 88−89
Climate change, 36, 137−138

Climate problem
    big actions on global climate issues, 84
    energy revolution and climate ambition,
        84−88
    energy revolution triggered by, 83−88
    nature, 83
Closed local system, 69
Closed-loop decision-making processes,
    54−57
Cloud computing, 75, 173−174
CloudPSS-IESLab, 212
Co-robot-driven smart manufacturing plant
    and process, 229−238
    cobots in age of Industry 5.0, 237−238
    future of cobot, 235−237
    human−robot collaboration, 229−234
CO₂ capture, use and storage (CCUS),
    16−18
Coal mine knowledge computing model,
    49−52
Coal mining industry, 49−52
Coal-fired power generation, 205
Cockpit, 225
Cognitive decision-making, 190−191
Cognitive machines, 20
Cognitive twin, 171, 189−190
    applications and challenges of, 195−196
    characteristics and structure of, 189−191
        autonomy capability, 190−191, 192f
        cognition capability, 189−190
        overall system lifecycle management,
            190
    function layers of, 193−195
        data management, 193
        model management, 194−195
        service management, 195
        six-layer progression model of
            cognitive twin, 194f
        twin management, 195
    vision, 193
Collaboration
    in HRC, 231f
    level of, 230−231
Collaborative industrial robots, 234
Collaborative robots (cobots), 25,
    150−151, 229−230, 233
    in age of Industry 5.0, 237−238

applications of, 151*f*

benefits of collaborative robots automation, 231–232, 232*f*

cognition process, 229, 230*f*

future of, 235–237, 236*f*

for industrial Internet-of-Things, 147–156

in Internet-of-Things, 149–152

smart industry system, 152–156

Combined heat, power, and fuel (CHPF), 131

Combined heat and power system (CHP), 126–127

Communication technologies, 220–221

Computing nodes, 145–147

Concrete product information, 266–267

Concurrent information, 176

Connectivity, 159

Construction of urban brain, 259

Consumer humanities network, 57–58

Controller area network (CAN), 221–222

Conventional industrial robots, 232

Cooperation, 281

Core value flow guidance, 285–286

COVID-19 pandemic, 114

CPS. *See* Cyber-physical systems (CPS)

Cross-disciplinary communication, 263

Cross-disciplinary fields, 275

Cross-disciplinary model, 131

Cross-disciplinary project, 280

Cross-domain integration, 284

Cross-industry transformation of transportation industry, 106–107

4N4F based on optical hydrogen storage, 108*f*

basic characteristics of 4N4F integration, 107*f*

Currents, 128–129

Customers, 152–153

Cutting-edge Hybrid Twin concept, 186

Cyber hygiene, 10

Cyber security, 7–8

Cyber threats to energy security, 8

Cyber world, 175, 283

directly transformation of domain knowledge into, 197

Cyber-attacks, 9–11

countermeasures to cyber-attacks on energy sector, 9–11

target operational technology, 8

Cyber-physical system-driven digital twin, 46, 119

Cyber-physical systems (CPS), 30, 151–152, 155–156

Cybersecurity attacks on energy sector, 9, 9*f*

Cyberspace, 28–29

D

Data acquisition, 135–136

Data fusion, 163–164

Data intelligence, 77

Data layer, 158

Data management, 193

Data mining techniques, 193

Data processing, 197–198

Data streams, 219–220

Data-based dynamic models, 183

Data-driven cyber-physical digital twin, 171–180

Data-driven digital twin concept, 173

Data-driven models, 171, 183

Data-driven service, 195

Data-information-knowledge intelligence, 133

Decarbonizing power sector, 5–6

Decision makers, 163

Decision-making, 3–4, 100, 113, 185, 238

capability, 207

process, 31

strategy, 175–176

Dedicated short-range communication (DSRC), 221–222

Deep decarbonization, 5–6

Deep integration technology system, 47

Deep learning (DL), 2–3, 49–52

Demand-side users, 97

DESDs. *See* Distributed energy storage devices (DESDs)

Developers, 228

Developing models, 195, 223–225

Device-edge-cloud model, 62

DGI. *See* Distributed grid intelligence
    (DGI)
Digital cockpit, 259
Digital computers, 62
Digital economy, 82—83
Digital energy
    industry, 49—52
    security, 9—11
Digital humanistic network, 39—40
Digital humanity network, 115, 120—121
    embodiment of humanity network in
        smart city, 119—120
        smart energy operating system, 118—119
        smart industry, 119
        smart transportation, 119
    in 4N4F, 115—120, 116f
Digital industry, 169—170
Digital infrastructure, 47—48, 279
Digital models, 189
Digital power grid, 206
Digital productivity, 64, 76, 88, 169—170,
    199
Digital renaissance, 196, 284
    data-driven cyber-physical digital twin,
        171—180
    design pattern from first principle,
        169—171, 170f
    intelligence-driven human cyber-
        physical cognitive twin, 188—199
    knowledge-driven human cyber hybrid
        twin, 180—188
Digital renaissance-driven smart city,
    245—262
    cases and practices, 257—262
        basic infrastructure, 258—259, 259f
        construction of urban brain, 259
        great activity to demonstration, 260
        livable cities for humanity, 261—262
        smart energy in building green
            ecological city, 260—261
        smart transportation in constructing
            hub city, 260
        top-level designs and policies, 258
        ubiquitous perception cities with
            information networks, 261, 262f
    emergence of smart city, 247—254
    smart city, 254—257

Digital social network, 21—22
Digital technologies, 4—5, 204—206, 254
Digital transformation of transportation
    industry, 101—106
    application practice of Shenzhen Smart
        Airport, 102f
Digital trend of modern and future city,
    252—254
Digital twins (DTs), 21, 24—25, 49—52,
    75, 170—171, 266—267, 269.
    See also Cognitive twin
    accurate data model in, 197
    application case study and challenge,
        177—180, 179t
    basic framework and elements in, 175f
    city, 254
    concept, 173—174, 174t
    digital twin-based system, 186—187
    model, 175—177, 177f
        integration, 196
        working flow in data models, 176f
Digital urban governance, 261—262
Digitalization, 197—198
Digitization, 180—181, 206
Digitizing European Industry Initiative
    (DEI), 249
Discrete systems, 181
Dispatching system, 260
Distributed energy
    collection devices, 125—126
    storage, 92, 206
Distributed energy storage devices
    (DESDs), 125—128
Distributed generation, 206
Distributed grid intelligence (DGI),
    127—128
Distributed Operation Design of Energy
    System, 132—133, 134f
Distributed power plants, 3—4
Distributed renewable energy resources
    (DRERs), 127—128
Distributed sources, 173
Distribution network planning of energy
    system, 133f
Distribution power system, 109
Distribution system operation (DSO), 92,
    132

DL. *See* Deep learning (DL)
Domain knowledge, 197
"Double carbon" standards, 47
DRERs. *See* Distributed renewable energy resources (DRERs)
Driving information display system, 225
DSO. *See* Distribution system operation (DSO)
DSRC. *See* Dedicated short-range communication (DSRC)
DTs. *See* Digital twins (DTs)
DWPT. *See* Dynamic wireless power transfer (DWPT)
Dynamic energy flow, 210−211
Dynamic knowledge, 193, 199
Dynamic models for processing and transmitting real-time information, 197−198
Dynamic perception, 141−142
Dynamic state estimation, 135−136
Dynamic wireless power transfer (DWPT), 140
Dystopia, 26−27

E

e-commerce sharing model, 88−89
Eagle Eye assistance system, 260
ECM. *See* Energy circuit method (ECM)
Ecosystem
    operating system for smart city, 156−165
    in smart city, 156−159
        subcategories of safety in industrial collaborative robotics, 152*f*
        system architecture of smart factory, 153*f*
Edge computing, 206−208
EI. *See* Energy internet (EI)
Electric cars, 213
Electric drive system, 213−214, 216−217
Electric energy, 215
Electric hydrogen production, 208−209
Electric internet of things, 206−208
Electric loads, 125−126
Electric motors, 215
Electric power systems, 52−53, 126−127
Electric vehicle drive system, 215−217, 217*f*

Electric vehicle supply equipment (EVSE), 128
Electric vehicles (EVs), 84−88, 126−127, 140, 186, 206, 212−213, 215−220
    battery management system, 105
    electrification, 215−217
    intelligentization, 218−220
    smart charging, 128
Electrical system, 130
Electricity, 22, 31−32, 128−129, 160−161, 205−206, 240−242, 271
    consumption, 6, 206, 208
    market, 206
    networks, 133−135
Electricity-gas-cooling-heating-steam, 211−212, 243−244
Electrification, 42, 67−69, 84−88, 102, 137−141, 213−217
    electric vehicle drive system, 215−217
    process, 137−138, 140
    transformation, 214−215, 217
Electrified energy system, 5−6
Electrocatalytic reduction, 13−14
Electronic control unit processors, 220−221
Electronic devices, 218−219
Electronic equipment, 226
Electronic gear shifting, 216−217
Electronic safety skin, 150
Electronic technology, 127−128, 220−221
Emerging technologies, 149
Emission reduction, 47−48, 243
End of life (EOL), 171, 190
End-side energy managers, 97
Endogenous basic framework, 42, 71, 118−119
Energy
    big data platforms, 96
    businesses, 239
    conservation, 47−48
    consumption, 96, 159
    conversion, 6−7
    delivery systems, 8
    digital economy, 47
    digitization, 48−49
    ecosystem, 11−14
        carbon neutrality in, 16−18

Energy (*Continued*)
  impact model of human activities on
    biological cycle of carbon, 12*f*
  roadmap, 15–16
  energy-material-energy-material, 66
  energy-saving services, 242
  flow, 65–66, 94, 255
  generation model, 132
  green transportation networks, 54, 65,
    77
  gridization, 69–70
  grids, 8, 132
  industry, 204
  intelligence, 66
  management, 186, 259
  model, 265–266
  networks, 52–53, 205–206, 242–243,
    247–249
  resources, 204–205
  revolution, 60, 91
    cases of 4N4F in energy revolution,
      98–99
    climate problems, 83–99
    development of energy revolution
      under 4N4F, 93–97
    opportunities and challenges under
      energy revolution, 88–93
    triggered by climate problem, 83–88
  saving, 243
  sector, 9
    countermeasures cyber-attacks on to,
      9–11
    cybersecurity attacks on, 9
  for smart automobile charging, 128–130
  storage, 15, 92
    entropy coupling energy, 67–69
    station, 52
    technology, 48–49, 140
    units, 160–161
  supply, 3–4
  system, 67–68, 84–88, 133
  technology innovation process, 53, 90
  transmission process, 210–211,
    242–243
  transportation, and information
    behavior-driven smart vehicle and
      green transportation, 212–228
  co-robot-driven smart manufacturing
    plant and process, 229–238
  digital renaissance-driven smart city,
    245–262
  intelligent integrated energy services
    and society ecosystem, 238–245
  new type power system with new
    energy as main body, 204–212
  practice of four network four flows
    integration, 262–287
  utilization, 41–42, 277
Energy circuit method (ECM), 136, 137*f*
Energy digitalization
  carbon neutrality from energy
    ecosystem, 11–18
  emerging sign of supersmart Society 5.0,
    26–36
  energy security from artificial
    intelligence–driven energy, 1–11
  evolutionary direction of society,
    29–33
    Industry 4.0 and Society 5.0, 30
    smart cities around world, 32–33
    society with intensive knowledge,
      30–31
    zero carbon society, 31–32
  industrial revolution 4.0 to 5.0, 18–26
  realizing Society 5.0, 33–36
    habitat innovation, 33–34
    industry–academia–government
      collaboration, 35–36
    promoting regional revitalization, 35
  Society 5.0, 26–29, 28*f*
    human-centered society, 27–28
    merging cyberspace with physical
      space, 28–29, 29*f*
Energy internet (EI), 54–55, 125–126,
  239
  composition of, 126*f*
  energy flow and information flow for
    smart automobile charging,
    128–130
  initiative, 125–136
  integrated energy system, 133–136
  smart energy system, 130–133
  summary, 125–128
Energy Operating System framework, 71

Energy security
  from artificial intelligence—driven
      energy, 1—11
  cyber threats to, 8
Engineering modeling, 183—184
Engineering systems, 182—183
Entropy, 66—68, 274—275
EOL. *See* End of life (EOL)
Equipment modelling, 210—211,
      242—243
Evaluation models, 175—176, 178—179
EVs. *See* Electric vehicles (EVs)
EVSE. *See* Electric vehicle supply
      equipment (EVSE)
Exergy, 66—68
Expert systems, 2
Extended reality (XR), 78, 263—264

F

Failure Mode, and Effective Analysis
      (FMEA), 172—173
Fashion themes, 250
FBS framework. *See* Function-Behavior-
      Structure framework (FBS
      framework)
FCA. *See* Fiat Chrysler Automobiles (FCA)
FEA. *See* Finite Element Analysis (FEA)
Feasible foundation methods, 192
Fee-reducing services, 242
Fiat Chrysler Automobiles (FCA),
      105—106
Fifth-generation cellular network (5G),
      173—174
  contemporary development and
      application of, 78
    characteristics of Metaverse, 79*f*
    development trend and benefits of
      4N4F, 80*f*
    industrial ecological innovation based
      on integration, 79*f*
    industry-wide empowerment of
      autonomous driving mode, 78*f*
  information network, 62
  IoV, 107—108
  networks, 75, 96
  technology, 44, 78, 149
Financing layer, 158

Finite Element Analysis (FEA), 181—182
First Industrial Revolution, 247—249
Flexibility, 233—234
  demands, 240—241
  resources, 240—241
FMEA. *See* Failure Mode, and Effective
      Analysis (FMEA)
Fossil energy, 14, 84—88, 90, 131
  power generation, 84—88
Fossil fuels, 13—14
  vehicles, 97
Fossil power generation, 91
Four industrial revolutions, 250
Four network four flows (4N4F), 169,
      208—209, 245—247, 250
  digital humanity network in, 115—120,
      116*f*
    smart city, 119—120
    smart energy operating system,
      118—119
    smart industry, 119
    smart transportation, 119
  based on optical hydrogen storage, 108*f*
  changes in transportation industry,
      101—107
    cross-industry transformation of
      transportation industry, 106—107
    digital transformation of transportation
      industry, 101—106
  development of energy revolution
      under, 93—97
  future of energy revolution, 93—97,
      95*f*
  drives transformation from Industry 4.0
      to Industry 5.0, 120—122, 121*f*
  in energy revolution, 98—99
    distributed operation demonstration of
      carbon-balanced energy ecosystem,
      99*f*
    distributed operation design of energy
      ecosystem, 98*f*
  integration, 45*f*, 262—263, 283
    basic elements of, 61*f*
    and innovation, 254—257
    systems, 270
  practice, 283—286
    renaissance from data to value, 285

Four network four flows (4N4F)
    (*Continued*)
    renaissance from fragment to holism,
        284–285
    renaissance from zero to one, 284
    renaissance of questioning spirit,
        285–286
    practice of four network four flows
        integration, 262–287
    system method, 66, 94
    value system, 271–276, 274*f*
    visualization, 263–271, 264*f*, 265*f*
Four networks fusion model, 65,
    111–122
    digital humanity network and
        application in 4N4F, 115–120
    4N4F drives transformation from
        Industry 4.0 to Industry 5.0,
        120–122
    people-oriented Industry 5.0, 111–115
        definition and characteristics of
            Industry 5.0, 113–114
        Industry 5.0 and Metaverse, 114–115
        origin of Industry 5.0, 111–112
Four Networks Integration, 45–46
    approach, 70
    industry, 61–62
Four Networks supersystem, 132
Four-network convergence model, 54
4K track camera system, 260
Fourth industrial revolution (Industry 4.0),
    88–89, 112, 235–238, 283
    and artificial intelligence, 89*f*
    4N4F drives transformation from
        Industry 4.0 to Industry 5.0,
        120–122
Fractal energy network system, 66
Free fuel-based power network, 138
FREEDM. *See* Future Renewable Electric
    Energy Delivery and Management
    Systems (FREEDM)
Fuel cells, 208–209, 215
Function-Behavior-Structure framework
    (FBS framework), 172–173
Future Renewable Electric Energy
    Delivery and Management Systems
    (FREEDM), 127–128

G
Gallium nitride (GaN), 216–217
Gas grids, 130
Gas networks, 125–126
Gas resources, 49–52
Gas systems, 240–241
GEIDCO. *See* Global Energy
    Interconnection Development and
    Cooperation Organization
    (GEIDCO)
GHG. *See* Greenhouse gas (GHG)
Global carbon dioxide emissions, 84
Global carbon emissions, 84
Global carbon peak, 88
Global climate issues
    actions on, 84
    carbon peaking and carbon
        neutralization schedule of count,
        85*t*
Global Energy Interconnection
    Development and Cooperation
    Organization (GEIDCO),
    127–128
Global greenhouse gas emissions, 204
Global positioning system (GPS), 130
Global warming, 83, 113–114
Google Maps, 157
Governance layer, 158
Government decision-makers, 269
Government Ownership and Oversight of
    Artificial Intelligence Data Act, 82
GPS. *See* Global positioning system (GPS)
Gravity model, 251
Green chemical industry, 131
Green economy, 60–64
Green energy, 58–59
    sources, 160–161
    supply, 209–210
Green hydrogen, 13–14
Green indicators, 47–48
Green spaces, 269
Green sustainability, 114–115
Green transportation, 212–228
    development of smart vehicles, 220–228
    future trends of electric vehicles,
        215–220
Green travel, 137–141

Greenhouse gas (GHG), 137
    accumulation, 83
    emissions, 13, 34

H

Habitat innovation, 33–34, 35*f*
HCNG. *See* Hydrogen-enriched
        compressed natural gas (HCNG)
Hetero organization, 66
Heterogeneous cloud, 199
Heterogeneous energy flow systems,
        208–209
Heterogeneous energy networks,
        208–209, 240
Hierarchical decoupling, 199
High speed 5G network transmission,
        265–266
High-speed information transmission
        technology, 58–59
Higher power electronic products, 215–216
Holographic battlefield planning, 83
HRC. *See* Human–robot collaboration
        (HRC)
Human care, 282
Human intelligence, 169
Human knowledge, 184
Human network, 281–282
    in human-machine-thing, 59–60
Human thinking process, 49–52
Human-artificial intelligence-environment,
        47–57
    challenges of artificial intelligence in
        renaissance, 52–56
    industrial humanity in renaissance,
        56–57
Human-centered society, 27–28
Human-centered urban governance,
        277–278
Human-cyber physical framework, 45–46
Human-cyber-physical ecosystem, 118
Human-cyber-physical elements, 43–44
Human-cyber-physical systems, 39–40, 56,
        115, 155–156, 169, 196–197
    Four Networks and Four Flows
        integration based on, 45*f*
    new production relationship formed by,
        41*f*

smart industrial application based on
        evolution of, 46*f*
Human-in-loop hybrid twin, 197–198
Human-information-energy-transportation
        elements, 43–44
Human-information-energy-transportation
        network, 169
Human-information-physical system, 88
Human-machine-things, 57–64
    definition of, 58–59
    importance of human network in
        human-machine-thing, 59–60
    network integration, 60–63
Human-oriented cognitive intelligence, 41,
        115
Human–centered renaissance, 39–46
    concept of industrial humanity, 39–41
    development of humanities network,
        41–43
    integration applications of humanistic
        network, 45–46
    integration key of cyberphysical-social,
        43–45
    renaissance
        challenge, 47–57
        elements, 57–64
        foundation, 64–71
Human–computer interaction method,
        58–59, 225
Humanistic network, 39–40, 58–64, 198
    integration applications of, 45–46
Humanistic relations network, 252–253
Humanity, 57–58
    livable cities for, 261–262
Humanity network, 88, 116, 257
    changes in mobility patterns about by
        transportation revolution, 99–111
    changes in transportation industry
        brought about by 4N4F, 101–107
    transportation reform in contemporary
        era, 99–111
    development of, 41–43
    embodiment
        in smart city, 119–120
        in smart energy operating system,
            118–119
        in smart industry, 119

Humanity network (*Continued*)
  in smart transportation, 119
  energy revolution solves climate
    problems, 83–99
  people oriented for four networks for
    fusion, 111–122
  social behavior patterns under
    information revolution, 73–83
  specific changes about change in
    transport pattern, 107–111
    highly automated and intelligent
      transportation system, 110–111,
      111*f*
    highly interconnected transportation
      system, 107–110, 109*f*
Human–machine hybrid intelligence,
    54–55, 57, 116
Human–machine integration, 78
Human–machine intelligence integration,
    71, 118–119
Human–machine intelligent fusion, 96
Human–machine smart agent, 199
Human–robot collaboration (HRC),
    229–234
  benefits of collaborative robots
    automation, 231–232
  better human and machine interface,
    233
  collaborative robot cognition process,
    230*f*
  difference between cobot and industry
    robot, 230
  different levels of collaboration in, 231*f*
  flexibility, 233–234
  increased ROI, 232
  level of collaboration, 230–231
  safety and satisfaction, 234
Human–vehicle interaction assistance,
    225–226
Human–vehicle–environment
    coordinated system, 221–222
Hybrid computing power, 199
Hybrid planning system, 102
Hybrid twin, 171, 183–185. *See also*
    Virtual twin
  application of, 186–188
    electric vehicle, 186

wind twin project, 186–188
  concept and model of, 183–186, 184*f*,
    185*f*
  technology, 186
Hydrogen, 13–14
  charging, 107
  energy, 42, 67–69, 84–88, 92
  storage, 107
Hydrogen-enriched compressed natural gas
    (HCNG), 208–209
Hydrogenation
  network, 106
  technology, 106

I

IaaS. *See* Infrastructure as a service (IaaS)
ICT. *See* Information and communication
    technology (ICT)
ICV. *See* Intelligent connected vehicles
    (ICV)
IEMS. *See* Integrated energy management
    system (IEMS)
IES. *See* Integrated energy system (IES)
IFR. *See* International Federation of
    Robotics (IFR)
IGFC technology. *See* Integrated
    gasification fuel cell technology
    (IGFC technology)
IIES. *See* Intelligent integrated energy
    services (IIES)
IIoT. *See* Industrial Internet-of-Things
    (IIoT)
IIRA. *See* Industrial Internet Reference
    Architecture (IIRA)
Immersive educational learning, 83
In-road sensors, 141–143
In-vehicle artificial intelligence, 220–221
In-vehicle entertainment system, 227–228
In-vehicle sensors, 141–143
Incidents, 185
Independent service providers, 209–210
Industrial blockchain, 25
Industrial business models, 40
Industrial collaborative robotics, 151
Industrial cooperative robot systems,
    151–152
Industrial development, 48–49

Industrial ecological joint innovation
model, 44
Industrial humanities, 22
concept of, 39–41, 40f
in renaissance, 56–57, 57f
Industrial Humanities Network, 53–54,
57–59
Industrial Internet, 54–55, 154f
Industrial Internet Alliance, 148
Industrial Internet Reference Architecture
(IIRA), 148
Industrial Internet-of-Things (IIoT),
147–149
collaborative robots for, 147–156
intelligence connected transportation
based on electric vehicles, 144f
relationship between emerging
technologies and key requirements,
149t
Industrial joint innovation process, 42, 70
Industrial process, 46
Industrial revolution, 64, 283–284
industrial revolution 4.0 to 5.0, 18–26
Industry 1.0 to Industry 4.0, 18–20,
19f, 20f
Industry 5.0, 21–23, 24f
technology enablers of Industry 5.0,
23–26
Industrial robots, 149–150
installation, 233–234
Industrial technology methods, 22, 40,
56–58
Industry 1.0, 111–112
Industry 5.0, 112–115, 237–238
cobots in age of, 237–238
definition and characteristics of,
113–114
4N4F drives transformation from
Industry 4.0 to, 120–122
origin of, 111–112
Industry robot, difference between cobot
and, 230
Industry–academia–government
collaboration, 35–36
Information and communication
technology (ICT), 10, 156,
276–277

Information entropy, 130
Information flow, 249
for smart automobile charging, 128–130
Information infrastructure, 280
Information layer, 158
Information model, 179
Information networks, 65
and future Metaverse integration and
application of, 81–83
perception cities with, 261, 262f
Information revolution, 74
development history and process of
information revolution, 73–76
development of sixth information
revolution, 75
history of information revolution, 74
information Internet to Metaverse,
75–76
integration and application of
information network technology
and future Metaverse, 81–83
social behavior patterns under, 73–83
Information technologies, 73–74,
127–128, 180–181
Infrastructure as a service (IaaS), 23
Innovation
of four network four flows, 254–257
infrastructure layer, 159
layer, 157–158
productivity, 255
Innovative integration, 279
Innovative solutions, 11
Integrated energy consumption system, 206
Integrated energy management system
(IEMS), 135–136, 136f
Integrated energy system (IES), 133–136,
208–209, 238–240
energy circuit method description, 137f
IES-Plan, 212
interconnection and coordination of
energy sectors, 135f
modelling, 210–211
multienergy devices in, 241f
operation optimization, 211, 243
planning methods, 211
simulation, 210–211, 242–243

Integrated gasification fuel cell technology
    (IGFC technology), 131
Integration
    of four network four flows, 254–257
    methodology along product lifecycle,
        198
    process, 279
Intelligence connected transportation,
        137–147
    classification of sensors in vehicles,
        144–147
    electrification, enabling green travel,
        137–141
    intelligence connected transportation,
        141–144
    smart transportation system, 144–147
Intelligence technology, 65
Intelligence transportation, 141
Intelligence-connected transportation, 137,
        143
Intelligence-driven human cyber-physical
        cognitive twin, 188–199
    applications and challenges of cognitive
        twin, 195–196
    characteristics and structure of cognitive
        twin, 189–191
    cognitive engineering journey, 188f
    enabling technology for model
        management, 191–193
    function layers of cognitive twin,
        193–195
    future milestones, 196–199
        accurate data model in digital twin,
            197
        directly transformation of domain
            knowledge into cyber world, 197
        dynamic models for processing and
            transmitting real-time information,
            197–198
        fully autonomous and intelligent
            decision-making smart agent, 199
        integration methodology along
            product lifecycle, 198
Intelligence-driven model, 188
Intelligent cognitive engine, 119
Intelligent computer systems, 143–144
Intelligent computing, 119–120

application system of urban intelligent
        computing, 162f
    framework of smart city ecosystem, 158f
    smart city with, 161–165
Intelligent connected vehicles (ICV), 273
Intelligent decision-making, 49–52,
        206–208
    smart agent, 199
Intelligent design method, 173
Intelligent devices, 204–205
Intelligent digital industry, 199
Intelligent dispatching system, 49–52
Intelligent driving technology, 219–220
Intelligent energy management, 208–209
Intelligent energy operating system, 96
Intelligent human-cyber-physical system,
        197
Intelligent industrial system, 152–153
Intelligent integrated energy services (IIES),
        208–210, 238–245
    basic characteristics, 240–242
    basic theory and methods for, 210–211,
        242–243, 244f
    integrated energy system and,
        238–240
Intelligent network system, 78
Intelligent technologies, 164
Intelligent transportation, 110
    design, 219
    highly automated, 110–111, 111f
    mode, 100–101
    network, 251–252
Intelligentization, 218–220
Interaction process, 48–49, 90
Interactive collaboration system, 83
Interactive virtual information, 264–265
Interconnected energy systems, 4
Intergovernmental Panel on Climate
        Change (IPCC), 13, 83
Internal R&D test verification system, 217
International Data Corporation, 223
International Federation of Robotics (IFR),
        230–231
Internet, 261
    development of network, 76f
    information internet to Metaverse, next
        stage of, 75–76

technologies of contemporary
  information revolution, 76—81
contemporary development and
  application of 5G technology, 78
contemporary development and
  application of artificial intelligence
  technology, 76—77
development and application of future
  Metaverse technology, 79—81, 81f
Internet of Vehicles (IoV), 94—95, 145,
  214, 220—222, 222f
Internet-of-Things (IoT), 3—4, 21, 26—27,
  49—52, 60—64, 74, 144—145,
  173—174, 207
  collaborative robots in, 149—152
  core elements of new generation of
    smart transportation system, 145f
  three-tier industrial Internet-of-Things
    system architecture, 148f
Internet-of-Vehicles, 42, 67—69
InterPlanetary File System, 76
Intuitive displays, 266
IoT. See Internet-of-Things (IoT)
IoV. See Internet of Vehicles (IoV)
IPCC. See Intergovernmental Panel on
  Climate Change (IPCC)

J
Joint industrial innovation, 90

K
Kano model, 172—173
Key performance indicators (KPIs), 34
Knowledge graph, 193—195
Knowledge system, 284—285
Knowledge-driven human cyber hybrid
  twin, 180—188
  application of hybrid twin, 186—188
  benefits of digitalization, 180f
  concept and challenges of virtual twin,
    181—183
  concept and model of hybrid twin,
    183—186, 184f
Knowledge-driven hybrid twin, 196
KPIs. See Key performance indicators
  (KPIs)

L
L1 passive interaction, 226
L2 semiactive interaction, 226
L3 active interaction, 226—227
L4 customized interaction, 227
Law of scientific development, 284—285
Life-safety-related technology, 223—225
LIN. See Local interconnect network (LIN)
Lithium battery, 273
Load prediction, 135—136
Local interconnect network (LIN),
  221—222
Logic Theorist, 1—2
Low carbon smart city, 159—161
  industrial internet architecture based on
    knowledge integration, 154f
  infrastructure of, 161f
  smart city components, 156f
Low-carbon cities, 159—160
Low-carbon emissions, 212
Low-carbon energy flow, 255
Low-carbon energy supply, 209—210
Low-power wide area networks (LP-
  WAN), 33
LP-WAN. See Low-power wide area
  networks (LP-WAN)

M
Machine intelligence, 43—44, 57, 169,
  185—186, 189
Machine learning (ML), 2, 49—52, 54, 65,
  69, 120—121, 164—165, 236—237
Macro-cosmic digital model, 82
Management layer, 158
Manipulation process, 236—237
Manufacturing execution system, 152—153
Marine ecosystems, 11—12
Mass-energy equation, 42, 67—69, 94—95
Material flow, 42, 94
Mercedes Benz EQC online Showroom,
  104
Metaverse, 75—76, 79, 81—82, 114—115,
  263
  alliance, 262
  development and application of, 79—81
  information internet to, 75—76
Microgrid, 66, 92, 132

Microsensors, 206−208
Middle of Life (MOL), 171, 190
Mixed reality (MR), 78, 263
ML. *See* Machine learning (ML)
Mobile automation methods, 238
Mobile energy storage, 132
Mobile internet, 21−22, 56, 75
Mobile payment methods, 100−101
Mobile phone app, 77
Mobile transportation network, 66, 94
Mobility as Service, 144−145
Model learners, 181
Model management, 194−195
    enabling technology for, 191−193
    knowledge graph, 193
    ontology, 192−193
Model order reduction (MOR), 184
    MOR-based discretization approaches, 184
Model-driven service, 195
Modeling process, 170−171, 180
Modern service system, 208
Modular technologies, 164
MOL. *See* Middle of Life (MOL)
MOR. *See* Model order reduction (MOR)
Motion speed requirements, 231
MR. *See* Mixed reality (MR)
Multi-in-one drive system, 216−217
Multicollaborative operator model, 116−117
Multidisciplinary collaboration, 281
Multienergy flow, 240−241
Multiple cloud, 199
Multiple device systems, 271
Multiple energy flows, 208−209
Multiple integrated energy equipment, 244−245
Multiple intelligent systems, 276
Multiple transportation carriers, 145−147
Multiscale, 210

N
Nanogrid, 66
NASA. *See* National Aeronautics and Space Administration (NASA)

National Aeronautics and Space Administration (NASA), 173
National defense system, 83
National grid, 273
National Highway Transportation Safety Administration, 222−223
National Renewable Energy Laboratory (NREL), 212, 244−245
Natural ecosystem, 11−14, 12*f*
Natural gas, 126−127
    networks, 133−136
    power generation, 205
Natural language processing (NLP), 236−237
Natural segmentation, 256−257
Near field communication (NFC), 33, 221−222
Network access technology, 81
Network integration, 60−63, 61*f*
Network modelling, 242−243
Networking supply chain, 256
Networking transition of urban layout, 217*f*, 251−252
Neural network, 251−252
Neutral climate ambitions, 88
New building energy conservation technologies, 240
New energy generation, 207
New energy service system, 208−209, 238−239
New operation system of smart city, 256−257
New type power system with new energy as main body, 204−212
    basic theory and methods for intelligent integrated energy services, 210−211
    empowering the new type power system with digital power grid, 206−208
    intelligent integrated energy services, 208−210
    main features of new type power system, 205−206
    typical intelligent integrated energy services platforms, 211−212
Newton's law of universal gravitation, 251

Next-generation mobile computing platform, 44
NFC. *See* Near field communication (NFC)
NICE_energy operating system (NICE_EOS), 132–133
NICE_EOS. *See* NICE_energy operating system (NICE_EOS)
Nicer-Net, 132–133
Nitrogen oxide emission, 159–160
NLP. *See* Natural language processing (NLP)
Nonrenewable energy, 12–13
NREL. *See* National Renewable Energy Laboratory (NREL)

O

OADT. *See* Overarching ADT (OADT)
OEM. *See* Original equipment manufacturer (OEM)
Oil resources, 49–52
On-board entertainment system, 225
Online human–computer interaction, 118
Ontology, 192–195
Operation optimization, 211, 243
Operational technology (OT), 8
Optimal energy flow, 135–136
Optimal power supply, 218–219
Organic matter, 11
Original equipment manufacturer (OEM), 214
OT. *See* Operational technology (OT)
OTA technology. *See* Over-the-air technology (OTA technology)
Over-the-air technology (OTA technology), 102
Overarching ADT (OADT), 178

P

P2G technology. *See* Power-gas technology (P2G technology)
PaaS. *See* Platform as a service (PaaS)
Paris Climate Change Conference (PCCC), 12–13
PCCC. *See* Paris Climate Change Conference (PCCC)

People-information-energy-transportation, 117
People-oriented smart society, 121–122
People-vehicle-road-net-cloud, 110
People-vehicle-road-network-cloud, 147
Perceive-think-execute-evolve process, 199
Performance indicators, 199
PEVs. *See* Plug-in EVs (PEVs)
Philosophy–science–engineering, 64–71
  engineering practice, 69–71, 70*f*
  integration of knowledge intelligence and data intelligence, 65*f*
  interactively integrating philosophical, 65–66, 67*f*
  scientific theory, 67–69, 68*f*
Photosynthesis, 11–12
Photovoltaic power generation, 90, 92
Photovoltaic storage-hydrogen-charging, 98
Planning process, 244–245, 247*f*
Platform as a service (PaaS), 23
PLM. *See* Product Lifecycle Management (PLM)
Plug-in EVs (PEVs), 128
Policy layer, 158
Pollution, 271–272
Power companies, 206–208
Power consumption, 205–206
Power distribution, 205–206
  network, 205–206
Power generation, 92, 205–206
Power grid, 92, 125–126, 205–206, 256–257
  operation, 6
  system, 49–52
Power sector, 204
Power supply, 256–257
  strategy, 47–48, 110–111
Power system, 91, 204–205
  dispatching system, 49–52
  distribution network, 132
Power transmission, 205–206
Power-gas technology (P2G technology), 126–127
Product consumers or users, 266–267
Product lifecycle, 171–172
  integration methodology, 198
  phases, 172*f*

Product Lifecycle Management (PLM), 173
Prototyping, 178
Public services, 282

Q
QFD. *See* Quality Function Deployment
    (QFD)
QoL. *See* Quality of life (QoL)
Quality Function Deployment (QFD),
    172−173
Quality of life (QoL), 34
Queuing hydrogenation, 218−219

R
Radio-frequency identification (RFID),
    221−222
Regional energy dispatching, 92
Renewable energy, 36, 90, 100, 131,
    137−138
    consumption, 211, 240, 243
    generation, 6
    production, 84−88
    systems, 125−126
    technologies, 3−4
Renewable power generation, 91
RFID. *See* Radio-frequency identification
    (RFID)
Road transportation efficiency, 277
Roadside units (RSU), 221−222
Robot as Service, 45−46
Robots, 149−150, 230−231
RSU. *See* Roadside units (RSU)

S
SaaS. *See* Software as a service (SaaS)
SAE. *See* Society of Automotive Engineers
    (SAE)
Safety of Human−robot collaboration, 234
Sale management system, 152−153
Satellite remote sensing, 261
Satisfaction of Human−robot
    collaboration, 234
Scenario-centric as-service ecological
    model, 154
Scientific theory, 67−69, 68*f*

Second law of thermodynamics, 95
Secondary energy networks, 126−127
Security
    of artificial intelligence−driven energy,
        7−11
    countermeasures to cyber-attacks on
        energy sector, 9−11
    cyber threats to energy security, 8
    cybersecurity attacks on energy sector,
        9, 9*f*
    assessment and control, 135−136
    layer, 159
Self-circulation evolution innovation
    process, 71
Self-driving network, 78
Self-driving vehicles, 143−144
Self-organization, 66
Self-similar operating system, 70
Semantic models, 57
Semantic technologies, 189−191
Semantization, 120
Sensors, 33, 114, 141, 157, 186−187, 261
    classification of sensors in vehicles,
        144−147
    technologies, 229, 234
Service management, 195
Seven-in-one drive system, 216−217
Simulated data, 173
Simulation, 242−243
    core, 187
    techniques, 184
    time, 178−179
Simulation-based engineering sciences, 171
Single technology development model,
        101−103
Single-point technologies, 81−82
Six-layer progression model, 193
Smart automobile charging
    blueprint of smart automobile charging,
        129*f*
    energy flow and information flow for,
        128−130
    integration of flows between energy and
        information in smart charging, 130*f*
Smart Brain of Xiongan, 258
Smart cars, 213

Smart Cities and Communities European
    Innovation Partnership (SCCEIP),
    249
Smart City Demonstration Projects, 249
Smart city/cities, 32–33, 156–157,
    159–160, 254–257, 278, 285–286
  components, 156f
  construction, 257–258
  conventional processes, 158
  development, 262
  ecosystem operating system for,
    156–165
    ecosystem in smart city, 156–159
    low carbon smart city, 159–161
    smart city with intelligent computing,
      161–165
  embodiment of humanity network in,
    119–120
  emergence of, 247–254
    digital trend of modern and future
      city, 252–254
    4N4F and smart city, 248f
    industrial revolution and smart city in
      different stages, 251f
    networking transition of urban layout,
      251–252
  four flows, 255–256
  four networks, 256–257
  around world, 32–33
    architecture of smart city, 32f
Smart cockpits, 214, 225–228
  interaction system, 227–228
  levels of smart cockpit in smart vehicles,
    226f
Smart electric vehicles, 212
Smart electricity, 130
Smart energy, 132
  in building green ecological city,
    260–261
  development, 42
  industry, 15–16
  integration, 90
  operating system, 42, 69–71, 95,
    118–119
    embodiment of humanity network in,
      118–119
  system, 41, 130–133

distributed operation design of energy
    system, 134f
  distribution network planning of
    energy system, 133f
  driven by multidomain knowledge
    integration, 131f
  implementation of distributed of
    carbon neutral energy system, 134f
Smart grids, 127, 260–261
Smart Growth Theory, 255
Smart industry, 152–153
  embodiment of humanity network in,
    119
  system, 152–156
    applications of collaborative robots,
      151f
    autonomous driving mode for data
      center, 152f
    comparison of cobots and
      noncollaborative robots, 150f
Smart information visualization system,
    266–267
Smart microgrids, 160
Smart Nation Platform, 249
Smart operating systems, 265
Smart robotics, 21
Smart sensors, 137
Smart solutions, 245–247
Smart transportation, 119, 132, 147
  in constructing hub city, 260
  embodiment of humanity network in,
    119
  system, 144–147
    core elements of new generation of,
      145f
    global CO_2 emissions from transport,
      138f
    services, 144–145
Smart vehicles, 218–219
  autonomous driving, 222–225
  development, 220–228
  internet of vehicles, 220–222
  smart cockpit, 225–228
Social media data, 163
Society of Automotive Engineers (SAE),
    223
Software as a service (SaaS), 23

Software-defined-energy framework, 71
Sound social security system, 278
Space-smart vehicle-working space, 227
Spatial scale, 210
Spatial-temporal big data, 163—164
Static energy flow, 210—211
Static numerical models, 182—183
Steam engines, 30
Suitable for business, 280
Suitable for living, 280—282, 282f
Supercomputing Cloud, 258
Supervisory control, 135—136
Supply chain optimization, 4
Sustainable circular ecology, 94
Sustainable energy, 55—56, 205—206
Sustainable innovation ecology, 276
Sustainable postindustrial production, 22
Sustainable power plants, 22
Sustainable recycling ecology, 66
Symbolism, 1—2
Synchronized vector approach, 128—129
Synergy, 208—209
System architects, 148
System integrators, 233—234
System lifecycle management, 189—190

T

Tail number specific ADT (TADT), 178
Technological innovations, 284
Technological productivity, 254—255
Technology infrastructure layer, 159
Telecommunication networks, 5—6
Terminal energy electrification, 60
Tesla's Model, 102
Theory of Inventive Problem Solving
    (TRIZ), 172—173
Thermal energy, 126—127
Thermal grids, 130
Thermal power generation, 84
Third Industrial Revolution, 127—128,
    249
3D model, 178—179
Three-in-one drive system, 216—217
Three-tier model, 148—149
Top-level designs and policies, 258
Traditional artificial intelligence, 117
Traditional consumer Internet, 53—54

Traditional design methods, 172
Traditional energy
    fields, 52
    services, 239
    system, 3—4
Traditional industrial robots, 229
Traditional operation monitoring, 207
Traditional rigid power system, 92
Traffic congestion, 277
Traffic indication system, 218—219
Traffic information, 268
Traffic management, 143
Transient process, 136
Transmission network technologies, 33
Transportation
    electrification, 140—141
    network, 247—249
    reform in contemporary era, 99—111,
        101f
Travel as Service system, 100—101
TRIZ. See Theory of Inventive Problem
    Solving (TRIZ)
Twin management, 195
    layer, 195

U

UNFCCC. See United Nations
    Framework Convention on
    Climate Change (UNFCCC)
United Nations Framework Convention
    on Climate Change (UNFCCC),
    12—13
Urban brain, construction of, 259
Urban data
    analysis, 164
    management, 163—164
Urban development, 256
Urban diseases, 245—247, 251
Urban Information Model, 258
Urban intelligent computing, 161,
    164—165
Urban intelligent development, 256—257
Urban model, 256—257
Urban transformation, 279
Urban-Rural Construction, 259
Urbanization, 283—284
Utilization of ecosystem, 159

V

V2G. *See* Vehicle-to-Grid (V2G)
V2X system, 273
Value evaluation system, 276
Value-added services, 208
VCU. *See* Vehicle computing unit (VCU)
Vehicle computing unit (VCU), 221–222
Vehicle electrification, 215
Vehicle-road network, 220–221
Vehicle-to-Grid (V2G), 138–140
    experimental project, 105
    networking services, 105–106
    technology, 128
Vehicle-to-Home, 138–140
Vehicle-to-Vehicle, 138–140
Vehicles, 223–225
Verbalization, 236–237
Virtual imaging technologies, 263–264
Virtual models, 189
Virtual power plant model, 97, 206, 208
Virtual reality (VR), 78, 151–152,
    249–250, 263
Virtual Space Industry, 82
Virtual twin, 181–182
    benefits, 182*f*
    concept and challenges of, 181–183
Vision technologies, 230

Visualization, 188–189, 266
    digital twin system, 269
Visualized integration system, 270
Voltages, 128–129
VR. *See* Virtual reality (VR)

W

Wastes, 11
Web 3.0, 76
Wind turbines, 187
Wind Twin project, 186–188
Wireless local area network (WLAN),
    221–222
Wireless sensor networks, 173–174
WLAN. *See* Wireless local area network
    (WLAN)

X

X as a Service (XaaS), 18–19
Xiongan smart city, 258
XR. *See* Extended reality (XR)

Z

Zero carbon society, 31–32
Zero-emission smart electric vehicles, 213

Printed in the United States
by Baker & Taylor Publisher Services